W9-DGE-262

Ergodic Theory and Topological Dynamics

Pure and Applied Mathematics

A Series of Monographs and Textbooks

RECENT TITLES

GERALD J. JANUSZ. Algebraic Number Fields

A. S. B. HOLLAND. Introduction to the Theory of Entire Functions

WAYNE ROBERTS AND DALE VARBERG. Convex Functions

A. M. OSTROWSKI. Solution of Equations in Euclidean and Banach Spaces, Third Edition of Solution of Equations and Systems of Equations

H. M. EDWARDS. Riemann's Zeta Function

SAMUEL EILENBERG. Automata, Languages, and Machines: Volumes A and B

MORRIS HIRSCH AND STEPHEN SMALE. Differential Equations, Dynamical Systems, and Linear Algebra

WILHELM MAGNUS. Noneuclidean Tesselations and Their Groups

FRANÇOIS TREVES. Basic Linear Partial Differential Equations

WILLIAM M. BOOTHBY. An Introduction to Differentiable Manifolds and Riemannian Geometry

BRAYTON GRAY. Homotopy Theory: An Introduction to Algebraic Topology

ROBERT A. ADAMS. Sobolev Spaces

JOHN J. BENEDETTO. Spectral Synthesis

D. V. WIDDER. The Heat Equation

IRVING EZRA SEGAL. Mathematical Cosmology and Extragalactic Astronomy

J. DIEUDONNÉ. Treatise on Analysis: Volume II, enlarged and corrected printing; Volume IV. *In preparation:* Volume V

WERNER GREUB, STEPHEN HALPERIN, AND RAY VANSTONE. Connections, Curvature, and Cohomology: Volume III, Cohomology of Principal Bundles and Homogeneous Spaces

I. MARTIN ISAACS. Character Theory of Finite Groups

JAMES R. BROWN. Ergodic Theory and Topological Dynamics

In preparation

CLIFFORD A. TRUESDELL. A First Course in Rational Continuum Mechanics: Volume 1, General Concepts

K. D. STROYAN AND W. A. J. LUXEMBURG. Introduction to the Theory of Infinitesimals

B. M. PUTTASWAMAIAH AND JOHN D. DIXON. Modular Representations of Finite Groups

MELVYN BERGER. Nonlinearity and Functional Analysis: Lectures on Nonlinear Problems in Mathematical Analysis

GEORGE GRATZER. Lattice Theory

Ergodic Theory and Topological Dynamics

JAMES R. BROWN
Department of Mathematics
Oregon State University
Corvallis, Oregon

ACADEMIC PRESS *New York San Francisco London* 1976
A Subsidiary of Harcourt Brace Jovanovich, Publishers

ACADEMIC PRESS, INC.
111 Fifth Avenue, New York, New York 10003

United Kingdom Edition published by
ACADEMIC PRESS, INC. (LONDON) LTD.
24/28 Oval Road, London NW1

Library of Congress Cataloging in Publication Data

Brown, James Russell, Date
Ergodic theory and topological dynamics.

(Pure and applied mathematics, a series of
monographs and textbooks ; v. 70)
1. Topological dynamics. 2. Ergodic theory.
I. Title. II. Series.
QA3.P8 [QA611.5] 510'.8s [515'.42] 75-40607
ISBN 0-12-137150-6

AMS (MOS) 1970 Subject Classifications: 28A65,54H20,
22D40,47A35

PRINTED IN THE UNITED STATES OF AMERICA

Jolan

Contents

Chapter III **Group Automorphisms and Affine Transformations**

Chapter IV **Entropy**

Chapter V **Bernoulli Systems and Ornstein's Theorem**

Preface

This book has been sixteen years aborning. In 1959–1960 the author sat in the lectures of S. Kakutani at Yale University and learned his first lessons in ergodic theory. Notes taken in those lectures have evolved and expanded over the years through the author's own lectures at Oregon State University, have been garnished by his contact with distinguished mentors and colleagues, and after numerous rewritings have assumed the form of the present text.

Why do we offer such a book now? Sixteen years ago in New Haven we were hearing of great new breakthroughs coming out of Moscow as the second great era of ergodic theory—the Kolmogorov era—was being launched. Now we are already six years into the third epoch—the Ornstein age. Perhaps the time has come to take a leisurely look at some of the accomplishments of the first two periods and a glimpse at what is developing in the third.

Chapters I and IV of this book are devoted, respectively, to pre- and post-Kolmogorov ergodic theory. Chapter V is an introduction to current developments. These three chapters may be read separately as an introduction to modern ergodic theory.

At the same time, the most casual observer must note the parallels of this theory to the topological dynamics introduced in Chapter II and Section 4 of Chapter IV. These parallels as well as the validity of two viewpoints, measure-theoretic and topological, in studying classical systems led the author to include the two half-brothers of ergodic theory and topological dynamics in one volume.

The two theories merge most satisfactorily when one studies the affine transformations of compact abelian groups, which are introduced in Chapter III.

No attempt at completeness has been made, and selections will invariably reflect the tastes of the author. Specific apologies are due, however, for the omission of the extensive body of results regarding invariant measures, the virtual groups of Mackey, and some important structure theorems of topological dynamics. The author reserves the privilege of returning to some of these as well as more specialized topics in a later volume.

It was intended that this book be accessible to the beginning mathematician with some background in abstract measure theory and general topology. In addition, a little familiarity with infinite-dimensional vector spaces and, for Chapter III, topological groups, would be helpful. It was further intended that the material would not be entirely without interest to the mature mathematician. We hope only that we have not quietly slipped into the interstice between these aims.

Exercises are included for all the chapters. While the number accompanying Chapter V is quite small, most readers will find the reading of that chapter sufficient exercise in itself.

CHAPTER

I

Ergodic Theory

1. ABSTRACT DYNAMICAL SYSTEMS

Ergodic theory may be defined to be the study of transformations or groups of transformations, which are defined on some measure space, which are measurable with respect to the measure structure of that space, and which leave invariant the measure of all measurable subsets of the space. In this chapter we shall concern ourselves with the theory of a single measure-preserving transformation and its iterates. This will make it possible to display the essential features of ergodic theory without becoming involved in unnecessary complications of notation and the intricacies of group theory. It should, however, be pointed out that most of the classical applications of ergodic theory require the consideration of a continuous group of transformations.

It is customary in ergodic theory to assume that the underlying space is either a finite or σ-finite measure space. We shall assume, except in some of the exercises, that the measure is finite and normalized to have total measure one. It is commonly further assumed that the measure space is *separable* (equivalently, that the space of square-integrable complex-valued functions on this measure space is a separable Hilbert space). We shall *not* make this assumption, principally because it would rule out some of our most interesting examples and our principal structure theorems in

Chapter III. There seems to be no compelling reason to impose the condition of separability, provided that we do our measure theory, in Chapter IV, for example, in terms of σ-algebras rather than partitions.

Let X be a nonempty set. Let $\mathscr{B}$ be a *σ-algebra* of subsets of X. In other words, $\mathscr{B}$ contains the empty set $\varnothing$ and the set X and is closed under the formation of countable unions, countable intersections, and complements. We make no further assumptions about $\mathscr{B}$. Let μ be a *normalized measure* on $(X, \mathscr{B})$. That is, μ is a nonnegative, real-valued, countably additive function defined on $\mathscr{B}$, with $\mu(X) = 1$. A function $\phi: X \to X$ is *measurable* if $\phi^{-1}(A) \in \mathscr{B}$ whenever $A \in \mathscr{B}$. Measurability for a function from one measure space to another is similarly defined. The measurable function $\phi: X \to X$ is said to be a *measure-preserving* transformation if $\mu(\phi^{-1}(A)) = \mu(A)$ for all $A \in \mathscr{B}$. It is an *invertible* measure-preserving transformation if it is one-to-one (monic) and if ϕ^{-1} is also measurable. In this case ϕ^{-1} is also a measure-preserving transformation.

Measure-preserving transformations arise, for example, in the study of classical dynamical systems. In this case ϕ is first obtained as a continuous transformation of some (compact) topological space, and the existence of an invariant measure μ, that is a measure preserved by ϕ, is proved. The system $(X, \mathscr{B}, \mu, \phi)$ is then abstracted from the topological setting. For this reason, we shall refer in this chapter to *abstract* dynamical systems.

Definition 1.1 An *abstract dynamical system* is a quadruple $\Phi = (X, \mathscr{B}, \mu, \phi)$, where X is a nonempty set, $\mathscr{B}$ is a σ-algebra of subsets of X, μ is a normalized measure defined on $\mathscr{B}$, and ϕ is a measure-preserving transformation of X. We shall say that Φ is *invertible* if ϕ is invertible.

While our principal object of study is, of course, the transformation ϕ, we adopt the above notation and terminology for a variety of reasons. For example, we shall have occasion to consider as different dynamical systems two quadruples $\Phi = (X, \mathscr{B}, \mu, \phi)$ and $\Phi' = (X, \mathscr{B}', \mu, \phi)$, which differ only in the class of measurable sets.

In order to avoid some trivialities as well as some embarrassing technical difficulties of measure theory, we shall adopt the following notion of equivalence of abstract dynamical systems.

Definition 1.2 The dynamical systems $\Phi = (X, \mathscr{B}, \mu, \phi)$ and $\Phi' = (X', \mathscr{B}', \mu', \phi')$ are *equivalent* if there exists a mapping $\psi^*: \mathscr{B}' \to \mathscr{B}$ which is monic and epic, and which satisfies

$$\mu(\psi^*(B')) = \mu'(B') \qquad (B' \in \mathscr{B}')$$

and

$$\mu[\phi^{-1}(\psi^*(B')) \,\Delta\, \psi^*(\phi^{-1}(B'))] = 0 \qquad (B' \in \mathscr{B}').$$

Here we have used the symbol Δ to denote the *symmetric* difference

$$C \,\Delta\, D = (C \sim D) \cup (D \sim C)$$

of the sets C and D.

Of course, if $\psi\colon X \to X'$ is an invertible measure-preserving transformation such that $\psi\phi = \phi'\psi$ modulo sets of measure zero, then its adjoint ψ^* defined by $\psi^*(B) = \psi^{-1}(B)$ satisfies the above requirements, so that Φ and Φ' are equivalent. However, it is not always possible to find such a ψ for equivalent systems Φ and Φ'.

Let us consider some examples of abstract dynamical systems.

Example 1 Let $X = [0, 1]$ be the set of real numbers between 0 and 1 inclusive. Let $\mathscr{B}$ denote the Borel subsets of X and μ the restriction of Lebesgue measure to X. Define $\phi(x)$ for each $x \in X$ to be the fractional part of $x + a$, where $0 < a < 1$. It is easily verified that ϕ preserves the length of any interval and hence (see Exercise 1) the measure of any set in $\mathscr{B}$.

Let $X' = K = \{z : |z| = 1\}$ be the set of complex numbers of absolute value one. Let $\mathscr{B}'$ denote the Borel subsets of X' and μ' the normalized linear measure on X'. Define $\phi'\colon X' \to X'$ by $\phi(z) = e^{2\pi i a}z$. Then $\Phi' = (X', \mathscr{B}', \mu', \phi')$ is a dynamical system, which is easily seen to be equivalent to $\Phi = (X, \mathscr{B}, \mu, \phi)$.

Let $X'' = X \times X$ be the unit square, and define $\phi''(x, y) = (\phi(x), y)$. If μ'' is two-dimensional Lebesgue measure, then $\Phi'' = (X'', \mathscr{B}'', \mu'', \phi'')$ will be equivalent to Φ provided that $\mathscr{B}'' = \{A \times X : A \in \mathscr{B}\}$, but *not* equivalent if $\mathscr{B}''$ is taken to be all the Borel subsets of X''.

Example 2 Let X, $\mathscr{B}$, and μ be as in the previous example, and define $\phi(x)$ to be the fractional part of kx, where k is a positive integer. Equivalently, define ϕ' on X' by $\phi'(z) = z^k$. This is an example of a non-invertible dynamical system (for $k > 1$). Since $\phi([0, 1/k]) = X$, it is not generally true for noninvertible systems that $\mu(\phi(A)) = \mu(A)$, even when $\phi(A)$ is measurable.

Example 3 (Shift transformations) Let

$$X_n = \{0, 1, \ldots, k-1\}$$

be a finite set of k points. Let $\mathscr{B}_n$ denote the class of all subsets of X_n, and let μ_n be the measure obtained by assigning to the point j the mass p_{nj}:

$$\mu_n = \{p_{n0}, p_{n1}, \ldots, p_{n,k-1}\}.$$

We can form a measure space $(X, \mathscr{B}, \mu)$ by taking the infinite product

$$(X, \mathscr{B}, \mu) = \mathop{\times}_{n=0}^{\infty} (X_n, \mathscr{B}_n, \mu_n)$$

or the two-sided product

$$(X', \mathscr{B}', \mu') = \mathop{\times}_{n=-\infty}^{\infty} (X_n, \mathscr{B}_n, \mu_n).$$

That is, X or X' consists of (one- or two-sided) sequences of elements of X_n, $\mathscr{B}$ or $\mathscr{B}'$ is the smallest σ-algebra containing all "cylinder sets"

$$C = \{x \in X : (x_{n_1}, \ldots, x_{n_l}) \in A\} = \bigcup_{(s_1, \ldots, s_l) \in A} \bigcap_{j=1}^{l} \{x : x_{n_j} = s_j\},$$

and μ or μ' assigns to C the measure

$$\mu(C) = \sum_{(s_1, \ldots, s_l) \in A} \prod_{j=1}^{l} p_{n_j, s_j}.$$

Now define ϕ and ϕ' on X and X', respectively, by $\phi(x) = y$ or $\phi'(x) = y$, where $y_n = x_{n+1}$ (all n).

Noting that

$$\begin{aligned}\phi^{-1}(C) &= \bigcup_{(s_1, \ldots, s_l) \in A} \bigcap_{j=1}^{l} \phi^{-1}\{y : y_{n_j} = s_j\} \\ &= \bigcup_{(s_1, \ldots, s_l) \in A} \bigcap_{j=1}^{l} \{x : x_{n_j+1} = s_j\},\end{aligned}$$

we see that ϕ^{-1} (or ϕ'^{-1}) carries cylinder sets into cylinder sets, hence is measurable. Clearly, it will be measure-preserving iff

$$p_{nj} = p_j \qquad (j = 0, 1, \ldots, k-1),$$

independently of n.

Note that ϕ' is invertible, but that ϕ is a k-to-one transformation. In fact, Φ is equivalent to the system Φ of Example 2. To see this, we need only express each $x \in [0, 1]$ in its k-adic expansion, thus obtaining an almost one-to-one correspondence between it and the one-sided sequence space X.

If $\Phi = (X, \mathscr{B}, \mu, \phi)$ is an abstract dynamical system, then ϕ determines a transformation T_ϕ of (real- or complex-valued) functions on X, defined by the formula $T_\phi(f(x)) = f(\phi(x))$. If f is measurable with respect to $\mathscr{B}$, then so is $T_\phi f$. If f is integrable, then $T_\phi f$ is integrable, and $\int T_\phi f \, d\mu = \int f \, d\mu$. This follows for simple functions from the fact that ϕ is measure preserving and in the general case by a limiting argument.

Recall that $L_p = L_p(\mu) = L_p(X, \mathscr{B}, \mu)$ $(1 \leq p < \infty)$ denotes the set of all measurable real- or complex-valued functions f defined on X for which $\int |f(x)|^p \mu(dx) < \infty$, and that L_p with the norm

$$\|f\|_p = \left[\int_X |f(x)|^p \mu(dx)\right]^{1/p}$$

is a complete, normed linear space. We denote, as usual, by L_∞ the space of μ-essentially bounded functions with the μ-essential supremum norm. In this chapter we shall be chiefly concerned with real L_p; that is, the functions in L_p will be assumed to be real valued. Note, however, that the ergodic theorems of Section 2 are valid in complex L_p, and that the spectral theory introduced in Section 3 requires consideration of complex L_p.

Since $T_\phi |f| = |T_\phi f|$, it is clear that the linear transformation T_ϕ maps L_p into L_p for each p, $1 \leq p \leq \infty$, and that $\|T_\phi f\|_p = \|f\|_p$. That is, T_ϕ is an isometry on L_p. If ϕ is an invertible measure-preserving transformation, then T_ϕ is an invertible isometry, with $T_\phi^{-1} = T_{\phi^{-1}}$. In the case $p = 2$, this means that T_ϕ is unitary.

We define a doubly stochastic operator on L_p $(1 \leq p \leq \infty)$, so called because of its origins in probability theory and its analogy to doubly stochastic matrices, as follows.

Definition 1.3 A linear operator T defined for functions on X is *doubly stochastic* if it maps L_1 into L_1 and satisfies for all $f \in L_1$ the following conditions:

1. $f \geq 0 \Rightarrow Tf \geq 0$;
2. $\int_X Tf \, d\mu = \int_X f \, d\mu$;
3. $T1 = 1$.

Here we use the symbols $\geq$ and $=$ in the μ-almost-everywhere sense, and we denote by 1 the function whose constant value is 1.

For each continuous linear operator T on L_p, there is a well-defined continuous linear operator T^*, called the *adjoint* of T, defined on L_q, where $1 \leq p < \infty$ and $1/p + 1/q = 1$ $(q = \infty$ for $p = 1)$. They are related by

$(Tf, g) = (f, T^*g)$ for all $f \in L_p$, $g \in L_q$. The symbol (f, g) denotes the "inner product"

$$(f, g) = \int_X f(x)\,\overline{g(x)}\,\mu(dx),$$

where $\bar{c}$ is the complex conjugate of c and the bar may be ignored for real L_p.

Proposition 1.1 *If T is a doubly stochastic operator, then T maps L_p into L_p for each p, $1 \le p \le \infty$, with $\|T\|_p \le 1$ and $\|T\|_1 = \|T\|_\infty = 1$. Moreover, T^* is also doubly stochastic.*

Proof We show first that $\|T\|_\infty = 1$. This makes sense because $L_\infty \subseteq L_1$, and T is defined on L_1. For $f \in L_\infty$, let f^+ and f^- denote, respectively, the positive and negative parts of f. Thus $f = f^+ - f^-$, $|f| = f^+ + f^-$, and f^+, $f^- \ge 0$. By property 1 of Definition 1.3, $Tf^+ \ge 0$ and $Tf^- \ge 0$. Thus

$$|Tf| = |T(f^+ - f^-)| = |Tf^+ - Tf^-| \le Tf^+ + Tf^- = T|f|. \tag{1}$$

By properties 1 and 3,

$$T|f| \le T\|f\|_\infty = \|f\|_\infty . \tag{2}$$

Combining inequalities (1) and (2) yields $\|Tf\|_\infty \le \|f\|_\infty$, or $\|T\|_\infty \le 1$. Since $T1 = 1$, it follows that $\|T\|_\infty = 1$.

Note that we used only properties 1 and 3 and the linearity of T to show that $\|T\|_\infty = 1$. Now property 2 is equivalent to saying that $(Tf, 1) = (f, T^*1) = (f, 1)$ for all $f \in L_1$. That is, property 2 is equivalent to $T^*1 = 1$. It follows as above that $\|T^*\|_\infty = \|T\|_1 = 1$. The inequality $\|T\|_p \le 1$ for $1 < p < \infty$ now follows as an application of the Riesz convexity theorem [16, p. 526]. (See also Exercise 7.)

To complete the proof we need to show that T^* maps L_1 into L_1 and satisfies condition 2. If $f \in L_\infty$, then $\int T^*f\,d\mu = (1, T^*f) = (T1, f) = (1, f) = \int f\,d\mu$. Since L_∞ is dense in L_1, it follows that T^* has a unique continuous extension to all of L_1 satisfying $\int T^*f\,d\mu = \int f\,d\mu$ for all $f \in L_1$. Integrating inequality (1) above with T replaced by T^* shows that T^* maps L_1 into L_1. ∎

It is interesting to note that the properties mentioned earlier for doubly stochastic operators arising from dynamical systems characterize these operators. Specifically, we have the following.

Proposition 1.2 *Suppose that* $(X, \mathscr{B}, \mu)$ *is the unit interval with Lebesgue measure on the Borel sets. Then the operators of the form* T_ϕ *for some dynamical system* $\Phi = (X, \mathscr{B}, \mu, \phi)$ *are just exactly the doubly stochastic operators which are isometries on* $L_2(X, \mathscr{B}, \mu)$. *Moreover,* Φ *is invertible iff* T_ϕ *is unitary.*

We shall not have occasion to use this result, and so will not give the proof. See, however, Exercise 6 at the end of this chapter.

2. ERGODIC THEOREMS

In the previous section we said that ergodic theory might be defined as the study of measure-preserving transformations. A more restrictive definition would be the study of the asymptotic behavior of the iterates ϕ^n of such a transformation. Indeed, the historical beginning of this discipline might be placed at the proof by G. D. Birkhoff in 1931 of the so-called individual ergodic theorem (Theorem 1.2) or the earlier proof by H. Poincaré in 1912 of the recurrence theorem (Theorem 1.5). In this section we shall look at these theorems as well as several others of a similar nature. We shall refer to these theorems collectively as ergodic theorems. Some of them involve the iterates ϕ^n of a measure-preserving transformation, while others involve the iterates T^n of an operator having some or all of the properties of Definition 1.3. We make no pretense at completeness or ultimate generality in our selection of ergodic theorems, but give only a representative sample of those we believe have had the most impact on ergodic theory and its applications.

One further historical note seems to be in order at this point. In retrospect it is clear that the mean and individual ergodic theorems for a measure-preserving transformation were anticipated considerably earlier by the (Weak) Law of Large Numbers of J. Bernoulli (1713) and the Strong Law of Large Numbers of E. Borel (1909) for Bernoulli sequences of random variables. The identification of these latter theorems as ergodic theorems only awaited the invention of measure theory by Borel and Lebesgue and its application by A. Kolmogorov in 1933 to the foundations of probability.

We begin with one of the so-called *maximal ergodic theorems*, this one due to E. Hopf. As before, let $(X, \mathscr{B}, \mu)$ be a normalized measure space. Let T be an operator on $L_1 = L_1(X, \mathscr{B}, \mu)$. We need only assume that T has property 1 of Definition 1.3 and a weakened form of 2, namely $\|T\|_1 \leq 1$. Such an operator is called a *contraction*. We introduce the following notation:

$$T_n f(x) = \sum_{k=0}^{n-1} T^k f(x),$$

$$B^*(f) = \{x : \sup_n T_n f(x) > 0\}, \qquad B_n^*(f) = \{x : \max_{1 \leq k \leq n} T_k f(x) > 0\},$$

where $f \in L_1$.

Theorem 1.1 (*Hopf maximal ergodic theorem*) *For each* $f \in L_1$

$$\int_{B^*(f)} f(x)\mu(dx) \geq 0. \tag{3}$$

Proof (A. Garsia [24]) Let

$$f_n(x) = \max_{1 \leq k \leq n} T_k f(x) \qquad (n = 1, 2, \ldots).$$

Then $f = f_1 \leq f_2 \leq \cdots$ and $B_n^* = B_n^*(f) = \{x : f_n(x) > 0\}$ is an increasing sequence of sets with union $B^*(f) = B^*$. Also, since $T \geq 0$,

$$T_1 f = f \leq f + Tf_n^+$$
$$T_{k+1} f = f + T(T_k f) \leq f + Tf_n^+ \qquad (1 \leq k \leq n)$$

so that

$$f_n \leq f_{n+1} \leq f + Tf_n^+$$

$$\int_{B_n^*} f\, d\mu \geq \int_{B_n^*} f_n\, d\mu - \int_{B_n^*} Tf_n^+\, d\mu$$
$$\geq \int_X f_n^+\, d\mu - \int_X Tf_n^+\, d\mu = \|f_n^+\|_1 - \|Tf_n^+\|_1 \geq 0.$$

The last inequality comes from the assumption that $\|T\|_1 \leq 1$. Letting $n \to \infty$, we obtain the desired result

$$\int_{B^*} f\, d\mu \geq 0. \quad \blacksquare$$

Now let us introduce the further notation:

$$A^*(f; \alpha) = \left\{x : \sup_n \frac{1}{n} T_n f(x) > \alpha\right\},$$

$$A_*(f; \alpha) = \left\{x : \inf_n \frac{1}{n} T_n f(x) < \alpha\right\}.$$

Corollary 1.1.1 *For each $f \in L_1$ and each real α and β we have*

$$\alpha\mu(A^*(f;\alpha)) \le \int_{A^*(f;\alpha)} f(x)\mu(dx) \tag{4}$$

and

$$\beta\mu(A_*(f;\beta)) \ge \int_{A_*(f;\beta)} f(x)\mu(dx). \tag{5}$$

Proof To prove (4) apply (3) to the function $h = f - \alpha \in L_1$ and observe that $T_n h(x) > 0$ iff $(1/n)T_n f(x) > \alpha$. Inequality (5) then follows by applying (4) to $g = -f$ and taking $\alpha = -\beta$. ∎

After particularizing this corollary to the operator T_ϕ induced by the dynamical system Φ, it is a relatively easy task to prove the most celebrated of the ergodic theorems, that of Birkhoff.

Theorem 1.2 (*Birkhoff individual ergodic theorem*) *Let $\Phi = (X, \mathscr{B}, \mu, \phi)$ be an abstract dynamical system, and suppose that $f \in L_1 = L_1(X, \mathscr{B}, \mu)$. Then there exists a function $\bar{f} \in L_1$ such that*

$$\lim_{n\to\infty} \frac{1}{n} \sum_{k=0}^{n-1} f(\phi^k(x)) = \bar{f}(x) \qquad \mu\text{-a.e.} \tag{6}$$

Proof (F. Riesz) Let us denote

$$\bar{f}(x) = \limsup_{n\to\infty} \frac{1}{n} T_n f(x), \qquad \underline{f}(x) = \liminf_{n\to\infty} \frac{1}{n} T_n f(x)$$

$$f^*(x) = \sup_n \frac{1}{n} T_n f(x), \qquad f_*(x) = \inf_n \frac{1}{n} T_n f(x),$$

so that in Corollary 1.1.1

$$A^*(f;\alpha) = \{x : f^*(x) > \alpha\}, \qquad A_*(f;\beta) = \{x : f_*(x) < \beta\}.$$

For fixed α and β with $\beta < \alpha$, let

$$A(\alpha, \beta) = \{x : \underline{f}(x) < \beta < \alpha < \bar{f}(x)\}. \tag{7}$$

Since $\bar{f}(\phi(x)) = \bar{f}(x)$ and $\underline{f}(\phi(x)) = \underline{f}(x)$ for all $x \in X$, it is clear that ϕ: $A(\alpha, \beta) \to A(\alpha, \beta)$. Assuming that $\mu(A(\alpha, \beta)) = \gamma > 0$, we can apply

Corollary 1.1.1 to the dynamical system $\Phi_{\alpha,\beta} = (A(\alpha, \beta), \mathscr{B} \cap A(\alpha, \beta), (1/\gamma)\mu, \phi)$. Since $f_* \leq \underline{f} \leq \bar{f} \leq f^*$, we have for $\Phi_{\alpha,\beta}$ that $A^*(f; \alpha) = A_*(f; \beta) = A(\alpha, \beta)$. It follows that

$$\alpha \leq \frac{1}{\gamma} \int_{A(\alpha,\beta)} f \, d\mu \leq \beta,$$

which contradicts $\beta < \alpha$. Thus we have $\mu(A(\alpha, \beta)) = 0$. Since

$$A = \{x : \underline{f}(x) < \bar{f}(x)\} = \bigcup_{\substack{\beta<\alpha; \\ \alpha,\,\beta \text{ rational}}} A(\alpha, \beta),$$

it follows that $\mu(A) = 0$. Thus $\underline{f}(x) = \bar{f}(x)$ μ-a.e., and the proof of convergence is complete.

To see that $\bar{f} \in L_1$, note that

$$\int_X \left| \frac{1}{n} T_n f(x) \right| \mu(dx) \leq \frac{1}{n} \int_X \sum_{k=0}^{n-1} |f(\phi^k(x))| \mu(dx) = \int_X |f(x)| \mu(dx).$$

By Fatou's lemma,

$$\int_X |\bar{f}| \, d\mu = \int_X \lim_{n\to\infty} \left| \frac{1}{n} T_n f \right| d\mu \leq \liminf_{n\to\infty} \int_X \left| \frac{1}{n} T_n f \right| d\mu$$

$$\leq \int_X |f| \, d\mu. \blacksquare$$

Remarks *1* Much has been done in the way of proving individual (that is, pointwise convergence) ergodic theorems for operators. See, for example, the excellent account by Garsia in [24]. A direct generalization of Theorem 1.2 to doubly stochastic operators yields the Hopf ergodic theorem. The proof again is based on Theorem 1.1. The same result with weaker hypotheses was proved by Dunford and Schwartz (see [16] or [24]). Recently, using the notion of "dilation of an operator," Akcoglu [3] has proved pointwise convergence for (positive) contractions on L_p, $1 < p < \infty$.

2 In the case of a discrete (completely atomic) measure space, a classical theorem of Kolmogorov yields convergence as in Eq. (6) for operators T only assumed to satisfy properties 1 and 3 of Definition 1.3. This theorem, usually stated in terms of convergence of a sequence of matrices, is basic in the analysis of finite or denumerable Markov chains.

3 Many of these theorems, including Theorem 1.2, are also valid when μ is a σ-finite measure. However, the limit function $\bar{f}$ may be uninteresting in this case (Exercise 9). A more sophisticated result, which also includes almost all of the theorems mentioned so far, is the following.

Theorem 1.3 (*R. Chacon–D. Ornstein*) *Suppose* $(X, \mathscr{B}, \mu)$ *is a finite or σ-finite measure space and* T *is a linear operator on* $L_1 = L_1(X, \mathscr{B}, \mu)$ *satisfying* (i) $T \geq 0$ *and* (ii) $\|T\|_1 \leq 1$. *Then for each* $f, g \in L_1$ *with* $g \geq 0$, *the limit*

$$\lim_{n\to\infty} \frac{T_n f(x)}{T_n g(x)}$$

exists and is finite almost everywhere that $\sup_n T_n g(x) > 0$.

The proof of this theorem is complicated and will not be given here (see [24, p. 30 ff.]). Instead, we proceed now to a fairly general "mean ergodic theorem," that is, one asserting convergence in L_p.

If Φ is a dynamical system, we shall see that for each $f \in L_p$ the sequence $(1/n)T_n f$ converges in the norm topology of L_p $(1 \leq p < \infty)$. It follows (Exercise 11) that the limit must coincide with $\bar{f}$ almost everywhere. Thus

$$\lim_{n\to\infty} \left\| \frac{1}{n} T_n f - \bar{f} \right\|_p = 0,$$

and, in particular, $\bar{f} \in L_p$.

In the following, we assume as before that $(X, \mathscr{B}, \mu)$ is a normalized, finite measure space.

Theorem 1.4 (*Yosida mean ergodic theorem*) *Let* T *be a doubly stochastic operator and* $f \in L_p$. *Then there exists* $f^* \in L_p$ *such that*

$$\lim_{n\to\infty} \left\| \frac{1}{n} \sum_{k=0}^{n-1} T^k f - f^* \right\|_p = 0. \tag{8}$$

Proof Suppose h is a function on X with $Th \leq h$. Then $g = h - Th \geq 0$ and $\int_X g \, d\mu = 0$. It follows that $g = 0$; that is, $Th = h$. According to Proposition 1.1, the same is true with T replaced by its adjoint T^*. In particular, if $T^*h_1 = h_1$ and $T^*h_2 = h_2$, then by the positivity of T^* and the previous remark, $T^*(h_1 \wedge h_2) = h_1 \wedge h_2$, where $h_1 \wedge h_2$ denotes the infimum of h_1 and h_2, defined by

$$(h_1 \wedge h_2)(x) = \min(h_1(x), h_2(x)).$$

Suppose $f = g - Tg$ with $g \in L_p$. Then $T_n f = g - T^n g$. It follows that

$$\left\| \frac{1}{n} T_n f \right\|_p \leq \frac{2}{n} \|g\|_p \to 0$$

as $n \to \infty$. Thus (8) holds with $f^* = 0$ for all f in the subspace $\mathscr{X}_1 = \{g - Tg : g \in L_p\}$. Likewise, (8) holds for all elements of $\mathscr{X}_2 = \{f \in L_p : Tf = f\}$, with $f^* = f$. We shall show that $\mathscr{X}_1 + \mathscr{X}_2$ is dense in L_p. It will follow that (8) is valid for all $f \in L_p$. For if $f_k \to f$ in L_p and (8) is valid for each f_k, then

$$\begin{aligned}\left\|\frac{1}{n} T_n f - \frac{1}{m} T_m f\right\|_p &\leq \frac{1}{n} \sum_{j=0}^{n-1} \|T^j(f - f_k)\|_p \\ &\quad + \frac{1}{m} \sum_{j=0}^{m-1} \|T^j(f - f_k)\|_p + \left\|\frac{1}{n} T_n f_k - \frac{1}{m} T_m f_k\right\|_p \\ &\leq 2\|f - f_k\|_p + \left\|\frac{1}{n} T_n f_k - \frac{1}{m} T_m f_k\right\|_p,\end{aligned}$$

which implies that (8) holds also for f.

To prove that $\mathscr{X}_1 + \mathscr{X}_2$ is dense in L_p, we shall show that the only $F \in L_q = L_p^*$ $(1/p + 1/q = 1$ for $p > 1$, and $q = \infty$ for $p = 1)$, which is orthogonal to both $\mathscr{X}_1$ and $\mathscr{X}_2$, is the zero function. Suppose then that F is such a function. It follows that $(F, g - Tg) = (F, g) - (T^*F, g) = 0$ for all $g \in L_p$, and hence that $T^*F = F$.

Let c be a fixed real number, and set $A = \{x : F(x) > c\}$. For each $\varepsilon > 0$ we define

$$g_\varepsilon = (1/\varepsilon)[(c + \varepsilon) \wedge F - (c \wedge F)].$$

Then $0 \leq g_\varepsilon \leq 1$ and $g_\varepsilon \uparrow \chi_A$, the characteristic function of the set A, as $\varepsilon \downarrow 0$. It follows by the monotone convergence theorem that $T^*g_\varepsilon \uparrow T^*\chi_A$ (Exercise 8). On the other hand,

$$\begin{aligned}T^*g_\varepsilon &= (1/\varepsilon)[T^*((c + \varepsilon) \wedge F) - T^*(c \wedge F)] \\ &= (1/\varepsilon)[(c + \varepsilon) \wedge F - (c \wedge F)] = g_\varepsilon,\end{aligned}$$

since F, c, and $c + \varepsilon$ are all invariant functions for T^*. Thus

$$T^*\chi_A = \lim_{\varepsilon \to 0} T^*g_\varepsilon = \lim_{\varepsilon \to 0} g_\varepsilon = \chi_A. \tag{9}$$

We shall show, in fact, $T\chi_A = \chi_A$.

If $B \in \mathscr{B}$ is arbitrary, then by (9) and the positivity of T

$$T^*\chi_{A \cap B} \leq T^*\chi_A = \chi_A \qquad \text{and} \qquad T^*\chi_{A \cap B} \leq T^*\chi_B,$$

so that

$$T^*\chi_{A \cap B} = \chi_A T^*\chi_{A \cap B} \leq \chi_A T^*\chi_B. \tag{10}$$

Likewise,

$$T^*\chi_{\tilde{A} \cap B} \leq \chi_{\tilde{A}} T^*\chi_B$$

or

$$\begin{aligned} T^*((1 - \chi_A)\chi_B) &\leq (1 - \chi_A)T^*\chi_B \\ T^*\chi_B - T^*\chi_{A \cap B} &\leq T^*\chi_B - \chi_A T^*\chi_B \\ \chi_A T^*\chi_B &\leq T^*\chi_{A \cap B}. \end{aligned} \tag{11}$$

Combining (10) and (11) gives $T^*\chi_{A \cap B} = \chi_A T^*\chi_B$. Thus

$$\begin{aligned} (T\chi_A, \chi_B) = (\chi_A, T^*\chi_B) = (1, \chi_A T^*\chi_B) &= (1, T^*\chi_{A \cap B}) = (T1, \chi_{A \cap B}) \\ &= (1, \chi_{A \cap B}) = (\chi_A, \chi_B). \end{aligned}$$

Since B was arbitrary, it follows that $T\chi_A = \chi_A$, as asserted.

Now $\chi_A \in L_\infty \subseteq L_p$, and hence $\chi_A \in \mathscr{X}_2$. Since F is assumed orthogonal to $\mathscr{X}_2$, this means that

$$(\chi_A, F) = \int_{[F > c]} F \, d\mu = 0.$$

Since this is true for all real c, we must have $F(x) = 0$ a.e. ▮

We conclude this section with a third type of ergodic theorem, the *recurrence theorem* of Poincaré.

Theorem 1.5 (*Poincaré*) *Let $\Phi = (X, \mathscr{B}, \mu, \phi)$ be an abstract dynamical system, and let $A \in \mathscr{B}$. Then for almost every $x \in A$ there is a positive integer $n = n_A(x)$ such that $\phi^n(x) \in A$.*

Proof Let

$$B = A \sim \bigcup_{n=1}^{\infty} \{x \in A : \phi^n(x) \in A\} = \bigcap_{n=1}^{\infty} (A \sim \phi^{-n}(A)).$$

Since

$$\phi^{-m}(B) = \bigcap_{n=1}^{\infty} (\phi^{-m}(A) \sim \phi^{-(n+m)}(A)),$$

it is clear that the sets $B, \phi^{-1}(B), \phi^{-2}(B), \ldots$ are pairwise disjoint, measurable sets. Since $\mu(\phi^{-n}(B)) = \mu(B)$ for each n, and since $\mu(X) = 1$, it follows that $\mu(B) = 0$. ▮

3. ERGODICITY AND MIXING

So far we know very little about the limit function $\bar{f}$ in Theorem 1.2. We know (see Exercise 11) that it coincides a.e. with the function f^* of Theorem 1.4, and that $\int \bar{f}\, d\mu = \int f\, d\mu$. Of course, in some special cases we can completely identify $\bar{f}$. For example, if f is an invariant function, $Tf = f$, then $\bar{f} = f$. We know that when $f = g - Tg$ for some $g \in L_1$, we have $\bar{f} = 0$. There is one more situation in which we can completely identify $\bar{f}$. This is when the dynamical system Φ is *ergodic*.

Definition 1.4 The abstract dynamical system $\Phi = (X, \mathscr{B}, \mu, \phi)$ is *ergodic* if $\phi^{-1}(A) = A$, $A \in \mathscr{B}$, implies either $\mu(A) = 0$ or $\mu(\tilde{A}) = 0$. A doubly stochastic operator T is *ergodic* if $Tf = f$, $f \in L_1$, implies that f is essentially a constant function, that is, $f(x) = c$ a.e.

Proposition 1.3 *If Φ is ergodic, then the induced operator T_ϕ is ergodic.*

Proof Suppose $T_\phi f = f$. For each positive integer n and each integer k let

$$X(k, n) = \left\{x : \frac{k}{2^n} \le f(x) < \frac{k+1}{2^n}\right\}.$$

Since ϕ is ergodic and $\phi^{-1}(X(k, n)) = X(k, n)$, it follows that $\mu(X(k, n)) = 0$ or $\mu(X(k, n)) = 1$. For each n there must be exactly one k with $\mu(X(k, n)) = 1$. Denote it by $k(n)$. It follows that $X_0 = \bigcap_{n=0}^{\infty} X(k(n), n)$ has measure 1. Clearly then, there exists a constant c such that $X_0 = \{x : f(x) = c\}$. ∎

Proposition 1.4 *If Φ is ergodic, then*

$$\lim_{n\to\infty} \frac{1}{n} \sum_{k=0}^{n-1} f(\phi^k(x)) = \int_X f\, d\mu \qquad \text{a.e.} \tag{12}$$

for each $f \in L_1$, and

$$\lim_{n\to\infty} \frac{1}{n} \sum_{k=0}^{n-1} \mu(A \cap \phi^{-k}(B)) = \mu(A)\,\mu(B) \tag{13}$$

for each $A, B \in \mathscr{B}$. Conversely, if (12) holds for all $f \in L_1$, or if (13) holds for all $A, B \in \mathscr{B}$, then ϕ is ergodic.

Proof The validity of (12) follows from $T_\phi \bar{f} = \bar{f}$ and $\int \bar{f}\, d\mu = \int f\, d\mu$. If we set $f = \chi_B$, then $f(\phi^k(x)) = \chi_{\phi^{-k}(B)}(x)$. Integrating (12) over A and applying the bounded convergence theorem yields (13).

Since (12) implies (13), it only remains to show that the validity of (13) for all $A, B \in \mathscr{B}$ implies that Φ is ergodic. Suppose $\phi^{-1}(B) = B$ and set $A = \tilde{B}$. The left side of (16) is zero then, and so either $\mu(B) = 0$ or $\mu(\tilde{B}) = 0$. ∎

The equality (12) is very closely related to the origins of ergodic theory in statistical mechanics. If we think of the sequence $\phi^n(x)$ as unfolding in time, then (12) is a statement of the *ergodic hypothesis*, namely, that time averages (of integrable functions) coincide with space (or phase) averages. In probability theory, (12) provides the foundation for a method of estimating parameters for (ergodic) stationary processes.

The significance of equality (13) is related to the recurrence theorem of Poincaré (Theorem 1.5). The latter theorem implies that, for a set A of positive measure, almost every point of A returns to A infinitely often. It gives us no information, however, as to how many points of A return to A at the nth step of the process, or, more generally, how many points of A are in the measurable set B after n steps. The proper measure of this number is $\mu(A \cap \phi^{-n}(B))$. Equality (13) tells us that asymptotically this number is *on the average* for different values of n proportional to the sizes of A and B. It may in fact (Exercise 18) *never*, for a given value of n, be close to $\mu(A)\,\mu(B)$.

Intuition tell us that for certain processes we should, in fact, have $\mu(A \cap \phi^{-n}(B))$ converging to $\mu(A)\,\mu(B)$ as $n \to \infty$. When this is true, the process is said to be *mixing* (or strongly mixing).

Definition 1.5 The dynamical system Φ is (*strongly*) *mixing* if

$$\lim_{n\to\infty} \mu(A \cap \phi^{-n}(B)) = \mu(A)\,\mu(B) \qquad \text{for all} \quad A, B \in \mathscr{B}.$$

To borrow an illustrative example from Halmos [32], suppose that a mixture is made containing 90% gin and 10% vermouth. If the process of stirring the mixture is ergodic, then after sufficient stirring any portion of the container will contain on the average (with respect to the number of stirrings) about 10% vermouth. If the process is a mixing one, the amount of vermouth in the given portion will become and remain close to 10%.

Since molecular theory allows for occasional "accidents," such as the kitchen table that rises into the air because all of its molecules are moving in the same direction, we may want to consider a slight weakening of the notion of mixing, namely that after a large number of stirrings the amount of vermouth in the distinguished portion of the container will be close to 10% *except for rare occasions*. We shall say that a set J of positive integers has *density zero* if the number of elements in $J \cap \{1, 2, \ldots, n\}$ divided by n tends to 0 as $n \to \infty$.

Definition 1.6 The dynamical system Φ is *weakly mixing* if for each $A, B \in \mathscr{B}$

$$\lim_{n\to\infty,\, n\notin J} \mu(A \cap \phi^{-n}(B)) = \mu(A)\,\mu(B), \tag{14}$$

where J is a set of density zero, which may vary for different choices of A and B.

The following proposition shows that weak mixing lies logically between mixing and ergodicity.

Proposition 1.5 *Let Φ be an abstract dynamical system. Then the following are equivalent:*

(i) *Φ is weakly mixing;*

(ii) $\lim_{n\to\infty} \frac{1}{n}\sum_{k=0}^{n-1} |\mu(A \cap \phi^{-k}(B)) - \mu(A)\,\mu(B)| = 0 \; (A, B \in \mathscr{B})$;

(iii) $\lim_{n\to\infty} \frac{1}{n}\sum_{k=0}^{n-1} [\mu(A \cap \phi^{-k}(B))]^2 = \mu(A)^2\,\mu(B)^2 \; (A, B \in \mathscr{B})$;

(iv) *the dynamical system $\Phi^2 = (X \times X,\ \mathscr{B} \times \mathscr{B},\ \mu \times \mu,\ \phi \times \phi)$ is ergodic, where $(\phi \times \phi)(x, y) = (\phi(x), \phi(y))$;*

(v) *Φ^2 is weakly mixing.*

Proof For a bounded sequence $\{a_n\}$ let us write $a = *\text{-}\lim_{n\to\infty} a_n$ provided that $a = \lim_{n\to\infty,\, n\notin J} a_n$, where J has density zero. Then, in general,

$$a = *\text{-}\lim_{n\to\infty} a_n \quad \text{iff} \quad \lim_{n\to\infty} \frac{1}{n}\sum_{k=0}^{n-1} |a_k - a| = 0. \tag{15}$$

For suppose that $a = *\text{-}\lim_{n\to\infty} a_n$, with the exceptional set being J. If $|a_n| \le b$ for all n, then

$$\frac{1}{n}\sum_{k=0}^{n-1} |a_k - a| \le \frac{1}{n}\sum_{k=0,\, k\notin J}^{n-1} |a_k - a| + \frac{1}{n}(b + |a|)|J_n|,$$

where $|J_n|$ is the number of elements in $J \cap \{0, 1, \ldots, n-1\}$. Thus $\lim_{n\to\infty} (1/n) \sum_{k=0}^{n-1} |a_k - a| = 0$.

To prove the converse, note that $a = *\text{-}\lim_{n\to\infty} a_n$ iff $J(\varepsilon) = \{n : |a - a_n| \geq \varepsilon\}$ has density zero for each $\varepsilon > 0$. For if the latter holds, then there exists an increasing sequence of integers n_m $(m = 1, 2, \ldots)$ such that

$$n \geq n_m \Rightarrow |J_n(1/m)| < n/m.$$

Setting

$$J = \bigcup_{m=1}^{\infty} \left[J\left(\frac{1}{m}\right) \cap [n_m, \infty) \right],$$

we have for each m

$$n \geq n_m, \qquad n \notin J \Rightarrow |a - a_n| < 1/m$$

and

$$n_{m+1} > n \geq n_m \Rightarrow |J_n| \leq \left| \bigcup_{j=1}^{m} J_n\left(\frac{1}{j}\right) \right| = \left| J_n\left(\frac{1}{m}\right) \right| < \frac{n}{m}.$$

Now suppose that $a \neq *\text{-}\lim_{n\to\infty} a_n$. Then there exist $\varepsilon_1 > 0$ and $\varepsilon_2 > 0$ such that $|J_n(\varepsilon_1)| \geq n\varepsilon_2$ for all n. It follows that

$$\frac{1}{n} \sum_{k=0}^{n-1} |a_k - a| \geq \frac{1}{n} \sum_{k=0,\, k \in J(\varepsilon_1)}^{n-1} |a_k - a| \geq \frac{\varepsilon_1}{n} |J_n(\varepsilon_1)| \geq \varepsilon_1 \varepsilon_2,$$

and hence

$$\lim_{n\to\infty} \frac{1}{n} \sum_{k=0}^{n-1} |a_k - a| \neq 0.$$

This completes the proof of equivalence (15).

Clearly, then (i) and (ii) are equivalent. Also

$$*\text{-}\lim_{n\to\infty} \mu(A \cap \phi^{-n}(B)) = \mu(A)\,\mu(B)$$

iff

$$*\text{-}\lim_{n\to\infty} |\mu(A \cap \phi^{-n}(B)) - \mu(A)\,\mu(B)| = 0$$

iff

$$*\text{-}\lim_{n\to\infty} |\mu(A \cap \phi^{-n}(B)) - \mu(A)\,\mu(B)|^2 = 0$$

iff

$$\lim_{n\to\infty} \frac{1}{n} \sum_{k=0}^{n-1} |\mu(A \cap \phi^{-n}(B)) - \mu(A)\,\mu(B)|^2 = 0. \tag{16}$$

Now if Φ is ergodic, then

$$\lim_{n\to\infty} \frac{1}{n} \sum_{k=0}^{n-1} |\mu(A \cap \phi^{-n}(B)) - \mu(A)\,\mu(B)|^2$$

$$= \lim_{n\to\infty} \frac{1}{n} \sum_{k=0}^{n-1} [\mu(A \cap \phi^{-n}(B))^2 - 2\mu(A)\,\mu(B)\,\mu(A \cap \phi^{-n}(B))]$$

$$+ \mu(A)^2\,\mu(B)^2$$

$$= \lim_{n\to\infty} \frac{1}{n} \sum_{k=0}^{n-1} \mu(A \cap \phi^{-n}(B))^2 - \mu(A)^2\,\mu(B)^2,$$

so that in this case (ii) and (iii) are equivalent. However, either (ii) or (iii) *implies* that Φ is ergodic. Thus (ii) and (iii) are equivalent.

To show that (14) holds for all $A, B \in \mathscr{B} \times \mathscr{B}$, with ϕ replaced by $\phi \times \phi$, it is sufficient to show that it holds for measurable rectangles. Condition (14) then becomes

$$*\text{-}\lim_{n\to\infty} \mu(A \cap \phi^{-n}(B))\,\mu(C \cap \phi^{-n}(D)) = \mu(A)\,\mu(B)\,\mu(C)\,\mu(D)$$

$$(A, B, C, D \in \mathscr{B}). \qquad (17)$$

Since the union of two sets of density zero has density zero, (17) follows from (14). That is, (i) implies (v).

Since (iv) obviously follows from (v), it only remains to show that (iv) implies (iii). If Φ^2 is ergodic and $A, B \in \mathscr{B}$, then

$$\lim_{n\to\infty} \frac{1}{n} \sum_{k=0}^{n-1} \mu(A \cap \phi^{-k}(B))^2$$

$$= \lim_{n\to\infty} \frac{1}{n} \sum_{k=0}^{n-1} (\mu \times \mu)[(A \times A) \cap (\phi \times \phi)^{-k}(B \times B)]$$

$$= (\mu \times \mu)(A \times A)(\mu \times \mu)(B \times B) = \mu(A)^2\,\mu(B)^2,$$

as was to be shown. ∎

It is time now to discuss some of the spectral properties of the operator T_ϕ on L_2. For this purpose, we consider T_ϕ to be operating on complex L_2. If $f \in L_2$ is a nonzero, complex-valued function such that $Tf = \lambda f$ for some complex number λ, we say that λ is an *eigenvalue* and f an *eigenfunction* of T. The collection of all eigenvalues of T is called the *point spectrum* of T. If $\mathscr{X}$ is a T-invariant subspace ($T\mathscr{X} = \mathscr{X}$) of L_2 containing no eigenfunctions of T, we say that T has *continuous spectrum* on $\mathscr{X}$. An eigenvalue λ of T is *simple* if $Tf = \lambda f$, $Tg = \lambda g$ implies that g is a constant multiple of f. If the invariant subspace $\mathscr{X}$ has a basis consisting of functions f_{ij} $(i = 1, 2, 3, \ldots; j = 0, \pm 1, \pm 2, \ldots)$ with $Tf_{ij} = f_{i,j+1}$ for each i and j, we say that T has *countable Lebesgue spectrum* on $\mathscr{X}$.

Theorem 1.6 *Let Φ be an invertible abstract dynamical system, and let T_ϕ be the induced operator on (complex) L_2. Then 1 is an eigenvalue of T_ϕ, and all eigenvalues have absolute value 1.*

(i) *If Φ is ergodic, then all eigenvalues are simple, and they form a subgroup of the multiplicative group $K = \{z : |z| = 1\}$.*

(ii) *If Φ is weakly mixing, then T_ϕ has continuous spectrum on the complement of the space of constant functions.*

(iii) *If T_ϕ has countable Lebesgue spectrum on the complement of the space of constant functions, then Φ is strongly mixing.*

Remarks *1* Constants are eigenfunctions corresponding to the eigenvalue 1. Thus T_ϕ cannot have continuous spectrum or Lebesgue spectrum on any space containing constants. By the complement of the space of constant functions, we mean the uniquely defined space $\mathscr{X} \subseteq L_2$ such that $L_2 = \mathscr{X} + \{\text{constants}\}$ and every function in $\mathscr{X}$ is orthogonal to 1, that is,

$$\mathscr{X} = \left\{ f - \int_X f \, d\mu : f \in L_2 \right\}.$$

2 The condition in (ii) is necessary *and* sufficient for Φ to be weakly mixing. For a proof see, for example, [32, pp. 39ff].

Proof We have already remarked that the constant functions are invariant, and hence that 1 belongs to the point spectrum. Since T_ϕ is unitary, all of its eigenvalues have absolute value one. Alternatively, if $T_\phi f = \lambda f$, then $T_\phi |f| = |\lambda| |f|$, and since $\int T_\phi |f| \, d\mu = |\lambda| \int |f| \, d\mu = \int |f| \, d\mu \neq 0$, it follows that $|\lambda| = 1$.

(i) According to Definition 1.4 and Proposition 1.3, Φ is ergodic iff 1 is a simple eigenvalue for T_ϕ. If Φ is ergodic, and if $T_\phi f = \lambda f$, then $T_\phi |f| = |f|$, so that $|f|$ must be a constant. If, in addition, $T_\phi g = \lambda g$, then $T_\phi(f/g) = (f/g)$, so that f/g is a constant. Finally, if $T_\phi f = \lambda_1 f$ and $T_\phi g = \lambda_2 g$, then $T_\phi(f/g) = (\lambda_1/\lambda_2)(f/g)$, so that λ_1/λ_2 is an eigenvalue.

(ii) Suppose Φ is weakly mixing. Then Φ^2 is ergodic. Suppose $T_\phi f = \lambda f$, and let $g(x, y) = f(x) \overline{f(y)}$. Then

$$g(\phi(x), \phi(y)) = f(\phi(x)) \overline{f(\phi(y))} = \lambda\bar{\lambda} f(x) \overline{f(y)} = g(x, y).$$

Thus g must be a constant. Hence f is a constant and $\lambda = 1$.

(iii) If f is a constant, then $(T_\phi{}^n f, g) = (f, g) = (f, 1)(1, g)$ for all n and each $g \in L_2$. If $f = f_{ij}$, $g = f_{pq}$, then $(T_\phi{}^n f, g) = (f_{i, j+n}, f_{pq}) = 0$ for all sufficiently large n. Since the functions f_{ij} plus the constant function 1 form a basis for L_2, it follows that

$$\lim_{n \to \infty} (T_\phi{}^n f, g) = (f, 1)(1, g) \qquad (f, g \in L_2).$$

In particular, this is true when f and g are characteristic functions and so Φ is strongly mixing. ▮

Let us look again at the examples of Section 1.

Example 1 Suppose first that a is irrational. In this case, Φ is ergodic but not weakly mixing, hence not strongly mixing. Using the alternate description of Φ on $K = \{z : |z| = 1\}$, we see that

$$T_\phi f(z) = f(e^{2\pi i a} z).$$

If $f_n(z) = z^n$, then

$$T_\phi{}^n f_n(z) = e^{2\pi i n a} z^n = c^n f_n(z).$$

Thus f_n is an eigenfunction with eigenvalue $\lambda = c^n \neq 1$. According to Theorem 1.6, Φ is not weakly mixing. On the other hand, any function $f \in L_2$ can be expanded in a Fourier series:

$$f \sim \sum_{n=-\infty}^{\infty} a_n f_n,$$

which converges to f in L_2. Thus if $T_\phi f = f$, it follows that f also has the expansion

$$T_\phi f \sim \sum_{n=-\infty}^{\infty} a_n c^n f_n.$$

By the uniqueness of the Fourier coefficients, it follows that $a_n = a_n c^n$ for each n. This means that $a_n = 0$ for $n \neq 0$, and so f is a constant.

If a is rational, then $c^n = 1$ for some positive integer n. Thus T_ϕ has nonconstant eigenfunctions, and so Φ is not ergodic.

Example 2 This system is equivalent to the system Φ of Example 3. This equivalence clearly preserves all properties of ergodicity and mixing. Note also that there is an induced unitary equivalence of the corresponding L_2 spaces. The system Φ is not invertible, so Theorem 1.6 does not apply.

Example 3 Φ' is strongly mixing. This follows from Theorem 1.6 by taking

$$f_{pq}(x) = e^{2\pi i p x_q / k}. \tag{18}$$

This example is a special case of a theorem about automorphisms of groups to be proved in Chapter III. It is fairly clear, and it will follow from a theorem on inverse limits, that the strong mixing property for Φ is equivalent to the same property for Φ'.

We have not shown yet that there exist systems Φ that are weakly mixing but not strongly mixing. This is a surprisingly difficult task, especially since it is now known that "most" systems are of this type. An example is given at the end of Section 6 and in the exercises.

4. PRODUCTS AND FACTORS

We begin now the study of methods for constructing new dynamical systems from given ones. This will lead in later chapters to representation theorems, whereby we express more complicated systems in terms of simpler, more familiar ones. The first such construction is the direct product $\Phi \otimes \Omega$ of dynamical systems Φ and Ω. We have already used a special case of this construction, namely $\Phi \otimes \Phi = \Phi^2$, in Proposition 1.5.

Definition 1.7 We define the *direct product* $\Phi_1 \otimes \Phi_2$ of abstract dynamical systems $\Phi_i = (X_i, \mathscr{B}_i, \mu_i, \phi_i)$ $(i = 1, 2)$ by

$$\Phi_1 \otimes \Phi_2 = (X_1 \times X_2, \mathscr{B}_1 \times \mathscr{B}_2, \mu_1 \times \mu_2, \phi_1 \times \phi_2),$$

where $(\phi_1 \times \phi_2)(x_1, x_2) = (\phi_1(x_1), \phi_2(x_2))$. More generally, if $\Phi_\alpha = (X_\alpha, \mathscr{B}_\alpha, \mu_\alpha, \phi_\alpha)$ is an abstract dynamical system for each $\alpha \in J$, we define the *direct product* $\Phi = \bigotimes_{\alpha \in J} \Phi_\alpha$ by taking the product measure structure on the product space $X = \mathsf{X}_{\alpha \in J} X_\alpha$ and defining

$$\phi(x) = y, \qquad \text{where} \quad y_\alpha = \phi_\alpha(x_\alpha). \tag{19}$$

We shall make use of customary modifications of this notation, such as $\Phi_1 \otimes \Phi_2 \otimes \cdots \otimes \Phi_n$ and $\bigotimes_{n=1}^{\infty} \Phi_n$.

Proposition 1.6 *The product of a weakly mixing system and an ergodic system is ergodic. The product of two weakly (strongly) mixing systems is weakly (strongly) mixing.*

Proof Let Φ_1 and Φ_2 be the two systems. It suffices to prove that (13) or (14) or the defining relation for strong mixing holds for pairs of measurable rectangles. That is, we need to show for all $A, B \in \mathscr{B}_1$ and $C, D \in \mathscr{B}_2$ that

$$\lim_{n\to\infty} \frac{1}{n} \sum_{k=0}^{n-1} \mu_1(A \cap \phi_1^{-k}(B))\, \mu_2(C \cap \phi_2^{-k}(D)) = \mu_1(A)\, \mu_1(B)\, \mu_2(C)\, \mu_2(D) \tag{20}$$

or

$$*\text{-}\lim_{n\to\infty} \mu_1(A \cap \phi_1^{-n}(B))\, \mu_2(C \cap \phi_2^{-n}(D)) = \mu_1(A)\, \mu_1(B)\, \mu_2(C)\, \mu_2(D) \tag{21}$$

or

$$\lim_{n\to\infty} \mu_1(A \cap \phi_1^{-n}(B))\, \mu_2(C \cap \phi_2^{-n}(D)) = \mu_1(A)\, \mu_1(B)\, \mu_2(C)\, \mu_2(D), \tag{22}$$

where Φ_1 is weakly mixing and Φ_2 is ergodic for (20), both are weakly mixing for (21), and both are strongly mixing for (22). The last one is completely obvious, while (21) depends only on knowing that the union of two sets of density zero has density zero. To prove (20) we note that, for a given $\varepsilon > 0$ and for all k larger than some $n_0 = n_0(\varepsilon)$, we may replace $\mu_1(A \cap \phi_1^{-k}(B))$ by $\mu_1(A)\, \mu_1(B) + \varepsilon_k$ with $|\varepsilon_k| < \varepsilon$, except when k belongs to some set J of density zero. Thus

$$\left| \frac{1}{n} \sum_{k=0}^{n-1} \mu_1(A \cap \phi_1^{-k}(B))\, \mu_2(C \cap \phi_2^{-k}(D)) - \frac{1}{n} \sum_{k=0}^{n-1} \mu_1(A)\, \mu_1(B)\, \mu_2(C \cap \phi_2^{-k}(D)) \right| \le n_0/n + ((n - n_0)/n)\varepsilon + (1/n)|J_n|.$$

The first and last terms on the right tend to zero as $n \to \infty$. Thus the two terms on the left have the same limit, namely the right side of (20). ∎

We shall see in the next section that the product of any number of weakly (strongly) mixing systems is weakly (strongly) mixing. Note, however, that the union of countably many sets of density zero need not have density zero.

Suppose that $\Phi = \Phi_1 \otimes \Phi_2$. Define $\psi\colon X \to X_1$ by $\psi(x, y) = x$. It follows easily that $\psi\phi = \phi_1\psi$. That is, the diagram

$$\begin{array}{ccc} X & \xrightarrow{\phi} & X \\ \downarrow{\scriptstyle\psi} & & \downarrow{\scriptstyle\psi} \\ X_1 & \xrightarrow{\phi_1} & X_1 \end{array} \tag{23}$$

commutes. Moreover, ψ is measure-preserving.

It is possible to have systems Φ and Φ_1 related by a map $\psi: X \to X_1$ for which diagram (23) commutes without Φ being of the form $\Phi_1 \otimes \Phi_2$.

Definition 1.8 We shall say that the dynamical system $\Phi_1 = (X_1, \mathscr{B}_1, \mu_1, \phi_1)$ is a *factor* of the system $\Phi = (X, \mathscr{B}, \mu, \phi)$ if there exists a measure-preserving map $\psi: X \to X_1$ such that diagram (23) commutes. In this case, we write $\Phi_1 | \Phi$ and $\psi: \Phi \to \Phi_1$ or $\Phi \overset{\psi}{\to} \Phi_1$. The map ψ is called a *homomorphism* of Φ onto Φ_1. If $\Phi = \Phi_1 \otimes \Phi_2$, Φ_1 is called a *direct factor* of Φ.

Note As usual in this chapter, when we write $\psi\phi = \phi_1\psi$, or indicate it by a diagram like (23), we mean that equality holds pointwise almost everywhere.

Suppose that $\Phi_1 | \Phi$. Let $\mathscr{B}_1' = \{\psi^{-1}(A) \colon A \in \mathscr{B}_1\}$. Then $\mathscr{B}_1' \subseteq \mathscr{B}$, and according to Definition 1.2, the systems Φ_1 and $\Phi_1' = (X, \mathscr{B}_1', \mu, \phi)$ are equivalent. Thus we may always assume that the factors of Φ are of this latter form. That is, the factors Φ_1 of Φ may be identified with the sub-σ-algebras $\mathscr{B}_1$ of $\mathscr{B}$ which are *invariant*, in the sense that $\phi^{-1}(\mathscr{B}_1) = \{\phi^{-1}(B) \colon B \in \mathscr{B}_1\} \subseteq \mathscr{B}_1$. Note that the factor Φ_1 is an *invertible* system iff $\mathscr{B}_1$ is *totally invariant*, that is, $\phi^{-1}(\mathscr{B}_1) = \mathscr{B}_1$ (Exercise 22).

It might be imagined that two dynamical systems Φ_1 and Φ_2 for which $\Phi_1 | \Phi_2$ and $\Phi_2 | \Phi_1$ are isomorphic, hence equivalent in the sense of Definition 1.2. However, it is not known if this is true even in the case where ϕ_1 and ϕ_2 are the identity; that is, the problem is unsolved even for measure spaces. Since the condition $\phi_1\psi = \psi\phi_2$ gives us a further restriction on the map ψ, the conjecture might conceivably be false for measure spaces, but true for ergodic dynamical systems, for example. This also is unsettled. It has become customary when $\Phi_1 | \Phi_2$ and $\Phi_2 | \Phi_1$ to say that Φ_1 and Φ_2 are *weakly isomorphic.*

Example Let us continue with Example 3 of Section 1. Recall that Φ was the one-sided shift on k points, with $X = \times_{n=1}^{\infty} X_n$, and Φ' was the two-sided shift, with $X' = \times_{n=-\infty}^{\infty} X_n$. Thus Φ' is invertible, but Φ is not. Define $\psi: X' \to X$ by $\psi(\ldots, x_{-1}, x_0, x_1, \ldots) = (x_1, x_2, \ldots)$. If

$$C = \{x \in X : (x_{n_1}, \ldots, x_{n_l}) \in A\} \tag{24}$$

is an arbitrary cylinder set in X, then $\psi^{-1}(C)$ has the same description with X replaced by X'. Thus ψ is measurable and measure-preserving. Clearly $\phi\psi = \psi\phi'$, so that $\Phi|\Phi'$. In particular, it follows from the following proposition and the discussion at the end of the preceding section that Φ is strongly mixing.

Proposition 1.7 *Suppose that* $\Phi_1|\Phi$ *and that* Φ *is* (1) *ergodic,* (2) *weakly mixing, or* (3) *mixing. Then* Φ_1 *has the same property.*

Proof If we represent Φ_1 in the form $\Phi_1 = (X, \mathscr{B}_1, \mu, \phi)$, where $\mathscr{B}_1$ is an invariant sub-σ-algebra of $\mathscr{B}$, then the relation given in (13), (14), or Definition 1.5 is true for all A, $B \in \mathscr{B}$, hence, in particular, for all A, $B \in \mathscr{B}_1$. ∎

5. INVERSE LIMITS

The direct product of infinitely many dynamical systems may be thought of as a limit of finite products in a way which will become clear in the following. On the other hand, the slightly more general notion of inverse limit is also useful in the calculation of entropy (Chapter IV) and the analysis of complex dynamical systems. Rather than a constructive definition, which is possible for the inverse limit of a sequence, we shall give a categorical definition of inverse limit, thus avoiding temporarily some of the sticky problems of existence. That is, our definition will involve only homomorphisms between dynamical systems and the completion of certain commutative diagrams. We note in passing that the direct product could also have been defined categorically.

Recall that a set J is said to be *directed* if there is defined on J a relation $\prec$ such that (i) $\prec$ is a partial order on J, and (ii) for each pair α, $\beta \in J$ there is a $\gamma \in J$ such that $\alpha \prec \gamma$ and $\beta \prec \gamma$.

Definition 1.9 By an *inverse system* of abstract dynamical systems we shall mean a triple $(J, \Phi_\alpha, \psi_{\alpha\beta})$ such that J is a directed set; for each $\alpha \in J$, Φ_α is an abstract dynamical system; and for each pair α, $\beta \in J$ with $\alpha \prec \beta$, we have $\psi_{\alpha\beta} : \Phi_\beta \to \Phi_\alpha$. An *upper bound* for such a system is a dynamical system Φ with a set of homomorphisms $\rho_\alpha : \Phi \to \Phi_\alpha$ ($\alpha \in J$) such that for each α, $\beta \in J$ the diagram

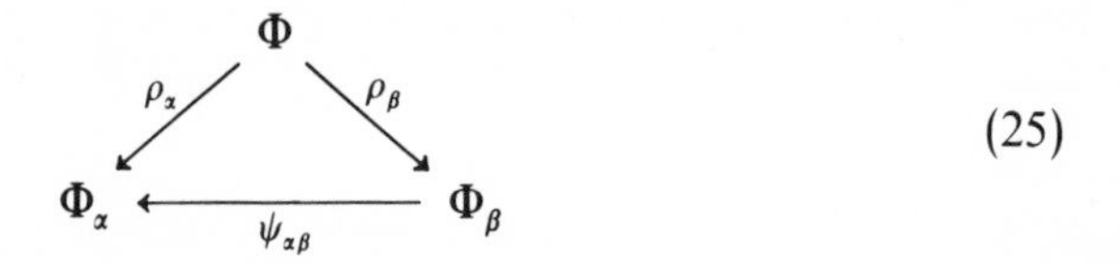

commutes. Finally, an *inverse limit* $\hat{\Phi}$ of the inverse system $(J, \Phi_\alpha, \psi_{\alpha\beta})$ is an upper bound with maps $\hat{\rho}_\alpha : \Phi \to \Phi_\alpha$ which is a factor of every other upper bound. That is, whenever Φ is an upper bound with maps $\rho_\alpha : \Phi \to \Phi_\alpha$, there exists a homomorphism $\sigma: \Phi \to \hat{\Phi}$ such that the diagram

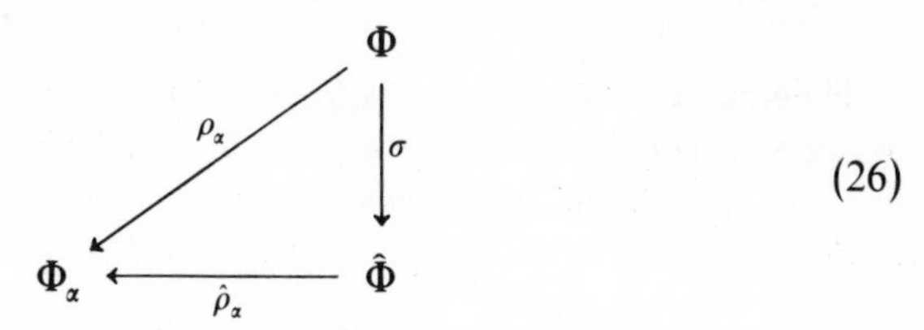

(26)

commutes for each $\alpha \in J$. In this case, we write

$$\hat{\Phi} = \operatorname*{inv\,lim}_{\alpha \in J} \Phi_\alpha \qquad \text{or} \qquad (\hat{\Phi}, \hat{\rho}_\alpha) = \operatorname*{inv\,lim}_{\alpha \in J} (\Phi_\alpha, \psi_{\alpha\beta}).$$

Clearly, if $\Phi = (X, \mathscr{B}, \mu, \phi)$ is an upper bound for the system $(J, \Phi_\alpha, \psi_{\alpha\beta})$, then we can represent the Φ_α as $(X, \mathscr{B}_\alpha, \mu, \phi)$, where the $\mathscr{B}_\alpha$ $(\alpha \in J)$ form an increasing net of invariant sub-σ-algebras of $\mathscr{B}$. The mappings $\psi_{\alpha\beta}$ and ρ_α then become the identity mapping on X. Moreover, the inverse limit $\hat{\Phi}$ can be identified with $(X, \hat{\mathscr{B}}, \mu, \phi)$, where $\hat{\mathscr{B}}$ is the smallest σ-algebra containing $\bigcup_{\alpha \in J} \mathscr{B}_\alpha$. This is true because $\hat{\mathscr{B}}$ reappears as an invariant sub-σ-algebra of $\mathscr{B}$ for *any* upper bound Φ, so that $\hat{\Phi} = (X, \hat{\mathscr{B}}, \mu, \phi)$ is a factor of Φ and the commutativity of (26) is trivial. In fact, this argument shows that any bounded system of abstract dynamical systems has an inverse limit, and that all such inverse limits are equivalent in the sense of Definition 1.2. Thus we have proved the following theorem.

Theorem 1.7 *If $(J, \Phi_\alpha, \psi_{\alpha\beta})$ is an inverse system of abstract dynamical systems, and if Φ is an upper bound with maps ρ_α, then*

$$\hat{\Phi} = (X, \hat{\mathscr{B}}, \mu, \phi) = \operatorname*{inv\,lim}_{\alpha \in J} \Phi_\alpha,$$

where $\hat{\mathscr{B}}$ is the smallest σ-algebra containing all of the $\rho_\alpha^{-1}(\mathscr{B}_\alpha)$. In particular, the inverse limit, when it exists, is uniquely determined up to equivalence.

The question of existence of the inverse limit is somewhat more difficult. The usual approach is to define the inverse limit set

$$X_\infty = \left\{ x \in \bigtimes_{\alpha \in J} X_\alpha : \psi_{\alpha\beta} x_\beta = x_\alpha \text{ for all } \alpha, \beta \in J, \quad \alpha \prec \beta \right\}, \tag{27}$$

define the projections $\rho_\alpha : X_\infty \to X_\alpha$ in the obvious way, and attempt to extend the measures $\mu_\alpha \rho_\alpha$ from $\mathscr{B}_0 = \bigcup_{\alpha \in J} \rho_\alpha^{-1}(\mathscr{B}_\alpha)$ to the σ-algebra $\mathscr{B}_\infty$

generated by $\mathscr{B}_0$. However, it is known (see, e.g., [31, p. 214]) that this is not always possible. For the most part, we shall be interested in inverse limits only when we have an explicit representation. However, the following theorem is not without interest. The proof [12, 14] is omitted.

Theorem 1.8 *Let $(J, \Phi_\alpha, \psi_{\alpha\beta})$ be an inverse system of abstract dynamical systems. Then there is a system $(J, \Phi_\alpha', \psi'_{\alpha\beta})$ such that Φ_α' is equivalent to Φ_α $(\alpha \in J)$ under a set of equivalences which carry the $\psi_{\alpha\beta}$ into $\psi'_{\alpha\beta}$ and such that*

$$\Phi_\infty = \operatorname*{inv\,lim}_{\alpha \in J} \Phi_\alpha'$$

exists. Moreover, Φ_∞ is defined on the inverse limit set (27) for $(J, \Phi_\alpha', \psi'_{\alpha\beta})$.

We shall see several examples of inverse limits in Chapter III. (See also Exercise 23.) For now we consider only two simple, but important examples. The first and most obvious is the direct product defined in the preceding section. For this we let I be the set of finite subsets of J, directed by set inclusion. Then

$$\bigotimes_{\alpha \in J} \Phi_\alpha = \operatorname*{inv\,lim}_{(\alpha_1, \ldots, \alpha_n) \in I} \Phi_{\alpha_1} \otimes \Phi_{\alpha_2} \otimes \cdots \otimes \Phi_{\alpha_n}.$$

The maps $\hat{\rho}_\alpha$ and $\psi_{\alpha\beta}$ are the obvious "finite-dimensional" projections, and a routine verification shows that the appropriate diagrams commute. If Φ is any other upper bound, then the map $\sigma\colon \Phi \to \bigotimes_{\alpha \in J} \Phi_\alpha$, given by $\sigma(x) = (\rho_\alpha(x))_{\alpha \in J}$, completes diagram (26).

As an application of Theorem 1.8, we give the following construction due to Rohlin [51]. If Φ is a noninvertible dynamical system, it is possible to define an invertible system $\hat{\Phi}$, called the *natural extension* of Φ, such that Φ is a factor of $\hat{\Phi}$, and $\hat{\Phi}$ is a factor of any other invertible system of which Φ is a factor (see Proposition 1.9 below).

For each positive integer n let $\Phi_n = \Phi$ and $\psi_{nm} = \phi^{m-n}$ for $m > n$. This defines an inverse system indexed by the set J of positive integers. Let $\hat{\Phi} = \operatorname{inv\,lim}_{n \in J} \Phi_n = \operatorname{inv\,lim}_{n \to \infty} \Phi_n$. Taking $\hat{\Phi} = \Phi_\infty$ as in Theorem 1.8, and noting that we can write

$$X_\infty = \{x \in \bigtimes_{n=1}^{\infty} X_n : x_n = \phi(x_{n+1}) \text{ for each } n\},$$

we see that

$$\hat{\phi}(x_1, x_2, x_3, \ldots) = (\phi(x_1), \phi(x_2), \phi(x_3), \ldots) = (\phi(x_1), x_1, x_2, \ldots).$$

Thus $\hat{\phi}$ is one-to-one, and its inverse

$$\hat{\phi}^{-1}(x_1, x_2, x_3, \ldots) = (x_2, x_3, x_4, \ldots)$$

is also measurable. That is, $\hat{\Phi}$ is invertible. Of course, if Φ is invertible, then $\hat{\Phi}$ is isomorphic to Φ. In fact, ρ_1 is an isomorphism, since $x_{n+1} = \phi^{-n} x_1$ for $x \in \hat{X}$.

Proposition 1.8 *Let $(J, \Phi_\alpha, \psi_{\alpha\beta})$ be an inverse system of dynamical systems, and let $J_0 \subseteq J$ have the property that for each $\alpha \in J$ there is a $\beta \in J_0$ such that $\alpha \prec \beta$. Then $(J_0, \Phi_\alpha, \psi_{\alpha\beta})$ is an inverse system, and*

$$\operatorname*{inv\,lim}_{\alpha \in J} \Phi_\alpha = \operatorname*{inv\,lim}_{\alpha \in J_0} \Phi_\alpha .$$

Proof This follows from the corresponding property for σ-algebras. Thus, if $\Phi_\alpha = (X, \mathscr{B}_\alpha, \mu, \phi)$, it is clear that $\bigcup_{\alpha \in J} \mathscr{B}_\alpha = \bigcup_{\alpha \in J_0} \mathscr{B}_\alpha$, and the result follows from Theorem 1.7. ▮

The proofs of the following two propositions are routine verifications and will be omitted.

Proposition 1.9 *If $\tau_\alpha : \Sigma_\alpha \to \Phi_\alpha$ for each $\alpha \in J$, then*

$$\operatorname*{inv\,lim}_{\alpha \in J}(\Sigma_\alpha, \omega_{\alpha\beta}) \,|\, \operatorname*{inv\,lim}_{\alpha \in J}(\Phi_\alpha, \psi_{\alpha\beta}),$$

provided that the diagrams

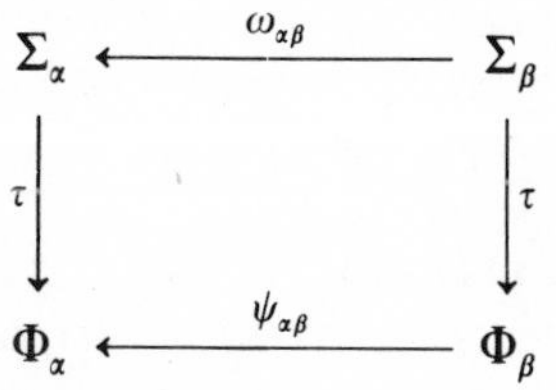

$$\begin{array}{ccc} \Sigma_\alpha & \xleftarrow{\ \omega_{\alpha\beta}\ } & \Sigma_\beta \\ \downarrow \tau & & \downarrow \tau \\ \Phi_\alpha & \xleftarrow{\ \psi_{\alpha\beta}\ } & \Phi_\beta \end{array}$$

commute for each α, $\beta \in J$. In particular, if $\Sigma | \Phi$, then $\hat{\Sigma} | \hat{\Phi}$, where the overcarat denotes the natural extension.

Suppose that Φ_1 and Φ_2 are factors of Φ. Then we may write $\Phi_k = (X, \mathscr{B}_k, \mu, \phi)$ $(k = 1, 2)$, where $\Phi = (X, \mathscr{B}, \mu, \phi)$. Let us denote by $\mathscr{B}_1 \vee \mathscr{B}_2$ the smallest σ-algebra containing both $\mathscr{B}_1$ and $\mathscr{B}_2$. We define the *join* of Φ_1 and Φ_2 to be $\Phi_1 \vee \Phi_2 = (X, \mathscr{B}_1 \vee \mathscr{B}_2, \mu, \phi)$. Of course, the notation and terminology extends to joins of arbitrary families of sub-σ-algebras of $\mathscr{B}$ and of factors of Φ.

Proposition 1.10 *If $\Phi_\alpha{}^1$ and $\Phi_\alpha{}^2$ are factors of Φ_α for each $\alpha \in J$, then*

$$\operatorname{inv}\lim_{\alpha \in J}(\Phi_\alpha{}^1 \vee \Phi_\alpha{}^2) = (\operatorname{inv}\lim_{\alpha \in J} \Phi_\alpha{}^1) \vee (\operatorname{inv}\lim_{\alpha \in J} \Phi_\alpha{}^2),$$

where the latter join is as factors of $\operatorname{inv}\lim_{\alpha \in J} \Phi_\alpha$. *In particular,*

$$\operatorname{inv}\lim_{\alpha \in J}(\Phi_\alpha{}^1 \otimes \Phi_\alpha{}^2) = (\operatorname{inv}\lim_{\alpha \in J} \Phi_\alpha{}^1) \otimes (\operatorname{inv}\lim_{\alpha \in J} \Phi_\alpha{}^2).$$

Proposition 1.11 *The inverse limit* $\operatorname{inv}\lim_{\alpha \in J} \Phi_\alpha = \Phi$ *is* (1) *ergodic,* (2) *weakly mixing, or* (3) *mixing iff each Φ_α has the same property.*

Proof Since each Φ_α is a factor of Φ, the result follows in one direction from Proposition 1.7. To prove the converse, let us denote $\Phi_\alpha = (X, \mathscr{B}_\alpha, \mu, \phi)$ where $\Phi = (X, \mathscr{B}, \mu, \phi)$. According to Theorem 1.7, the algebra $\mathscr{B}_0 = \bigcup_{\alpha \in J} \mathscr{B}_\alpha$ is dense in $\mathscr{B}$. Thus (see Exercise 19) condition (13) or (14) or the defining property of mixing holds on $\mathscr{B}$ iff it holds on $\mathscr{B}_0$. But the latter is true iff it holds on each $\mathscr{B}_\alpha$. ∎

Corollary *The natural extension of Φ is* (1) *ergodic,* (2) *weakly mixing, or* (3) *mixing iff Φ is.*

6. INDUCED SYSTEMS

In 1943, Kakutani [36] introduced the idea of a transformation induced by a measure-preserving transformation ϕ on a subset A of positive measure. The idea is to localize the system and only observe $\phi^n(x)$ when it is in A. This has been a very fruitful idea for constructing examples and has recently begun to play a role in the theory of abstract dynamical systems somewhat analogous to that of factor systems.

The basis of the construction is the recurrence theorem of Poincaré (Theorem 1.5). Thus if ϕ is a measure-preserving transformation on a finite measure space $(X, \mathscr{B}, \mu)$, and if $A \in \mathscr{B}$ is a measurable set of positive measure, then for almost every $x \in A$ there is a positive integer $n = n_A(x)$ such that $\phi^n(x) \in A$, but $\phi(x), \phi^2(x), \ldots, \phi^{n-1}(x) \notin A$.

Definition 1.10 The *induced transformation* on a set $A \in \mathscr{B}$ with $\mu(A) > 0$ is the transformation $\phi_A : A \to A$ defined by $\phi_A(x) = \phi^{n_A(x)}(x)$, where $n_A(x)$ is the smallest positive integer n such that $\phi^n(x) \in A$. The *induced dynamical system* is $\Phi_A = (A, \mathscr{B}_A, \mu_A, \phi_A)$, where $\mathscr{B}_A = \{A \cap B : B \in \mathscr{B}\}$ and μ_A is the normalized (total measure one) restriction of μ to $\mathscr{B}_A$.

Of course, ϕ_A is in general only defined for almost all $x \in A$. Its definition may be extended arbitrarily to all of A.

Theorem 1.9 *The induced transformation ϕ_A is measure preserving. Thus Φ_A is an abstract dynamical system. If Φ is invertible, so is Φ_A.*

Proof Define for $n \geq 1$

$$\begin{aligned} A_n &= \{x \in A : n_A(x) = n\} \\ &= \{x : x, \phi^n(x) \in A; \phi(x), \ldots, \phi^{n-1}(x) \notin A\} \\ B_n &= \{x : x, \phi(x), \ldots, \phi^{n-1}(x) \notin A; \phi^n(x) \in A\}. \end{aligned}$$

Since ϕ is measurable, we have A_n, $B_n \in \mathscr{B}$. Moreover, $A_n \subseteq A$ and for each $C \in \mathscr{B}$

$$\phi_A^{-1}(C) = \bigcup_{n=1}^{\infty} [A_n \cap \phi^{-n}(C)]. \tag{28}$$

It follows that ϕ_A is measurable on $(A, \mathscr{B}_A)$. Now the sets A_n $(n = 1, 2, \ldots)$ form a disjoint partition of (almost all of) A, and the sets B_n $(n = 1, 2, \ldots)$ form a disjoint partition of the set of all points whose "orbits" intersect A minus A (almost all of $X \sim A$ in the case of an ergodic Φ). Also

$$\begin{aligned} \phi^{-1}(A) &= A_1 \cup B_1 \\ \phi^{-1}(B_n) &= A_{n+1} \cup B_{n+1} \qquad (n \geq 1). \end{aligned} \tag{29}$$

For any $C \in \mathscr{B}$ with $C \subseteq A$, it follows by repeated application of (29) that

$$\begin{aligned} \phi^{-1}(C) &= [A_1 \cap \phi^{-1}(C)] \cup [B_1 \cap \phi^{-1}(C)] \\ \phi^{-1}[B_n \cap \phi^{-n}(C)] &= [A_{n+1} \cap \phi^{-(n+1)}(C)] \cup [B_{n+1} \cap \phi^{-(n+1)}(C)], \end{aligned}$$

or, since ϕ is measure preserving,

$$\mu(C) = \sum_{k=1}^{n} \mu(A_k \cap \phi^{-k}(C)) + \mu(B_n \cap \phi^{-n}(C)).$$

Since the B_n are pairwise disjoint, the last term tends to zero. Thus from (28)

$$\mu(C) = \sum_{n=1}^{\infty} \mu(A_n \cap \phi^{-n}(C)) = \mu(\phi_A^{-1}(C)).$$

Now suppose that Φ is invertible. Then, of course, ϕ^{-1} is measure preserving, and we can define $(\phi^{-1})_A$, the induced transformation on A. We shall show that $(\phi^{-1})_A = (\phi_A)^{-1}$. By symmetry it is sufficient to show that $(\phi^{-1})_A(\phi_A(x)) = x$ for almost all $x \in A$. Suppose that $x \in A_n$. Then $\phi_A(x) = \phi^n(x) = y \in A$. Clearly, $\phi^{-n}(y) = x \in A$. Suppose that $z = \phi^{-m}(y) \in A$ for some m, $1 \leq m < n$. Then $\phi^m(z) = y = \phi^m(\phi^{n-m}(x))$. But $\phi^{n-m}(x) \notin A$, and this contradicts the fact that ϕ^m is one-to-one. It follows that $(\phi^{-1})_A(y) = \phi^{-n}(y) = x$, as was to be shown. ▮

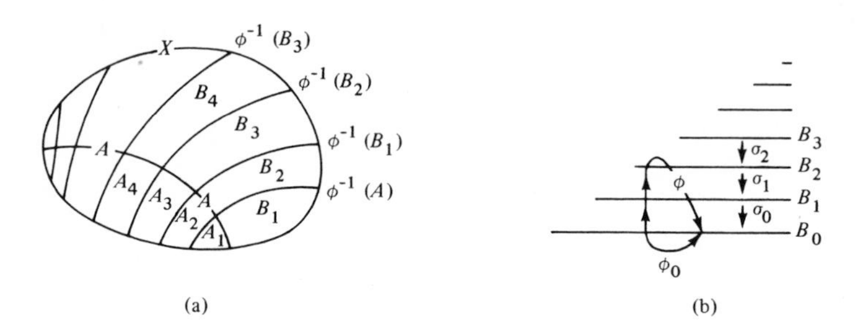

Figure 1. (a) Induced transformation; (b) inverse construction.

Proposition 1.12 *If Φ is ergodic, so is Φ_A.*

Proof Suppose that $C \in \mathscr{B}$, $C \subseteq A$, and $\phi_A^{-1}(C) \subseteq C$. Define

$$D = \bigcup_{n=1}^{\infty} \{[A_n \cap \phi^{-n}(C)] \cup [B_n \cap \phi^{-n}(C)]\}.$$

According to (28), $A_n \cap \phi^{-n}(C) \subseteq C$ for each n, and so $A \sim C \subseteq \tilde{D}$. On the other hand, by (29)

$$\phi^{-1}[A_n \cap \phi^{-n}(C)] \subseteq \phi^{-1}(C) = [A_1 \cap \phi^{-1}(C)] \cup [B_1 \cap \phi^{-1}(C)]$$

and

$$\phi^{-1}[B_n \cap \phi^{-n}(C)] = [A_{n+1} \cap \phi^{-(n+1)}(C)] \cup [B_{n+1} \cap \phi^{-(n+1)}(C)].$$

Thus $\phi^{-1}(D) \subseteq D$. Since Φ is ergodic, either $\mu(D) = 0$ or $\mu(\tilde{D}) = 0$. Suppose that $\mu(D) = 0$. It follows from (28) that $\mu(\phi_A^{-1}(C)) = 0$, and hence that $\mu(C) = 0$. Likewise, if $\mu(\tilde{D}) = 0$, then from the preceding $\mu(A \sim C) = \mu_A(\tilde{C}) = 0$. It follows that Φ_A is ergodic. ▮

In case Φ is invertible, there is an interesting way of describing the transformation ϕ in terms of ϕ_A and the sets B_n. This will lead to another new construction, which is, in a definite sense described below, the inverse of the induced transformation construction.

Let us write $B_0 = A$, so that $X = \bigcup_{n=0}^{\infty} B_n$. Note that ϕ maps B_{n+1} onto a subset of B_n for each n, and that ϕ^{-1} maps $x \in B_n \sim \phi(B_{n+1})$ onto the point $\phi_A^{-1}\phi^n(x)$.

Now suppose we are given a disjoint sequence of sets $B_n \in \mathscr{B}$, where $(X, \mathscr{B}, \mu)$ is a finite or σ-finite measure space. Suppose further that $\mu(B_{n+1}) \le \mu(B_n) < \infty$ and $\mu(B_n) \to 0$ as $n \to \infty$. For each n let $\sigma_n : B_{n+1} \to B_n$ be an invertible measure-preserving transformation of B_{n+1} onto $\sigma_n(B_{n+1})$. Let $\phi_0 : B_0 \to B_0$ be an invertible measure-preserving transformation of B_0 onto itself. We define a mapping ϕ: $Y \to Y$, where $Y = \bigcup_{n=0}^{\infty} B_n$, by

$$\begin{aligned} \phi(x) &= \sigma_n^{-1}(x) && \text{if} \quad x \in \sigma_n(B_{n+1}) \\ &= \phi_0 \sigma_0 \cdots \sigma_{n-1}(x) && \text{if} \quad x \in B_n \sim \sigma_n(B_{n+1}) \end{aligned} \qquad (n = 0, 1, \ldots).$$

Theorem 1.10 *The mapping ϕ is an invertible measure-preserving transformation. If $\mu(Y) = 1$, $\Phi = (Y, \mathscr{B}, \mu, \phi)$ is an abstract dynamical system, and ϕ_0 is the transformation induced by ϕ on B_0. If ϕ_0 is ergodic, so is ϕ.*

We leave the proof as an exercise.

Suppose Φ_1 and Φ_2 are invertible, ergodic dynamical systems. Let us write $\Phi_1 \prec \Phi_2$ if Φ_1 is isomorphic to a system Φ_{2A} induced by Φ_2 on some set A of positive measure. In Kakutani's terminology, Φ_1 is a *derivative* of Φ_2, and Φ_2 is a *primitive* of Φ_1. There is a clear analogy to the theory of factors discussed in Section 4, and again the question arises as to whether $\Phi_1 \prec \Phi_2$ and $\Phi_2 \prec \Phi_1$ imply $\Phi_1 \cong \Phi_2$.

The construction preceding Theorem 1.10 may be described by saying that $\Phi = (Y, \mathscr{B}, \mu, \phi)$ is *constructed on* the system $\Phi_0 = (B_0, \mathscr{B}_0, \mu_0, \phi_0)$. From the discussion it is clear that this is equivalent to $\Phi_0 \prec \Phi$, at least when Φ is ergodic. A discussion of this in terms of "flows under a function" is given in the exercises.

Example (*Kakutani* [37]) Let B_0 be the unit interval with Lebesgue measure for μ. (We can take, for example, $X = R \times Z$ to be the product of the reals with the integers.) Define ϕ_0 on B_0 by mapping the left half of the dyadic interval $[1/2^n, 1)$ linearly onto the right half:

$$\phi_0(x) = x - 1 + \frac{1}{2^n} + \frac{1}{2^{n+1}}, \qquad 1 - \frac{1}{2^n} \le x < 1 - \frac{1}{2^{n+1}}, \qquad n = 0, 1, 2, \ldots.$$

Let B_1 be a linear set of length $\frac{2}{3}$ "sitting above"

$$A = \bigcup_{n=0}^{\infty} \left[\frac{1}{2^{2n}}, \frac{1}{2^{2n+1}}\right),$$

and let $B_n = \varnothing$ $(n > 1)$. It is easily seen that ϕ_0, and therefore also ϕ, are ergodic. A little more effort (Exercise 35) reveals that Φ_0 has *discrete spectrum*, that is, T_{ϕ_0} has enough eigenfunctions to span $L_2(B_0)$. According to Theorem 1.6, Φ_0 is not weakly mixing. On the other hand, Φ is weakly mixing, but not strongly mixing (Exercise 36). Thus Proposition 1.12 fails if "ergodic" is replaced by "weakly mixing."

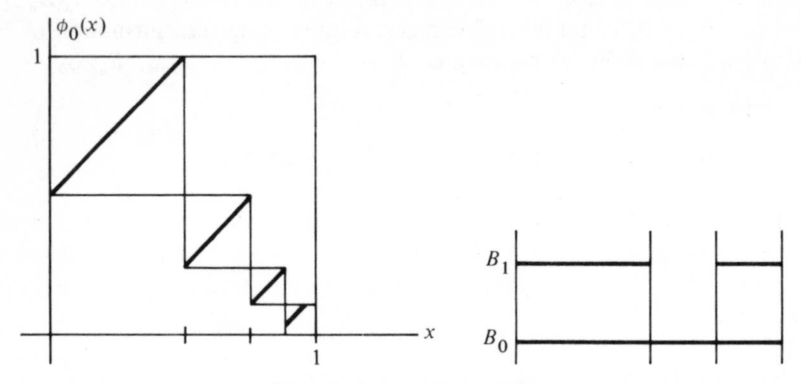

Figure 2. Kakutani's example.

EXERCISES

Measure-Preserving Transformations

1. **(a)** If $\mathscr{C}$ is a class of subsets of some set, let $\mathscr{A}(\mathscr{C})$ denote the smallest algebra of sets containing $\mathscr{C}$, and let $\mathscr{B}(\mathscr{C})$ denote the smallest σ-algebra containing $\mathscr{C}$. Suppose that $(X_i, \mathscr{B}_i, \mu_i)$ $(i = 1, 2)$ are finite measure spaces, and that $\phi\colon X_1 \to X_2$. If $\mathscr{C} \subseteq \mathscr{B}_2$ with $\mathscr{B}(\mathscr{C}) = \mathscr{B}_2$, and if $\phi^{-1}(B) \in \mathscr{B}_1$ for all $B \in \mathscr{C}$, show that ϕ is measurable.

(b) If, in addition, $\mathscr{C}$ satisfies

$$A, B \in \mathscr{C} \Rightarrow A \sim B \text{ is a finite union of pairwise disjoint sets in } \mathscr{C}. \tag{30}$$

and if $\mu_1(\phi^{-1}(C)) = \mu_2(C)$ for all $C \in \mathscr{C}$, then ϕ is measure preserving.

(c) The class $\mathscr{C}$ of measurable rectangles in a product space satisfies (30).

2. If $(X, \mathscr{B}, \mu)$ is a σ-finite measure space, we define measure-preserving transformations of X in exactly the same way as for a finite measure space. Does $\phi(x) = x + 2$ define a measure-preserving transformation of (i) the reals with Lebesgue measure, of (ii) the positive reals, of (iii) the integers with counting measure? How about $\phi(x) = 2x$? Show that $\phi(x, y) = (2x, y/2)$ is a measure-preserving transformation of the Euclidean plane.

3. (*Baker's transformation*) Define ϕ on the unit square $[0, 1] \times [0, 1]$ by $\phi(x, y) = (2x, y/2)$ for $0 \leq x < \frac{1}{2}$ and $\phi(x, y) = (2x - 1, (y + 1)/2)$ for $\frac{1}{2} \leq x \leq 1$.

(a) Show that ϕ is measure preserving.

(b) By mapping the sequence $\{x_n\}$ of 0's and 1's onto the point (x, y) such that x has the binary expansion $.x_0 x_1 x_2 \cdots$ and y the expansion $.x_{-1}x_{-2}x_{-3}\cdots$, show that ϕ is equivalent to the two-sided shift on two points.

4. Verify Example 2 and show that it is equivalent to the one-sided shift.

5. (*Adding machine transformation*) Define

$$(X, \mathscr{B}, \mu) = \mathop{\times}_{n=1}^{\infty} (X_n, \mathscr{B}_n, \mu_n),$$

where $X_n = \{0, 1, \ldots, k_n\}$, $\mathscr{B}_n$ is the class of all subsets of X_n, and $\mu_n = \{p_{n0}, p_{n1}, \ldots, p_{nk_n}\}$. Define $\phi\colon X \to X$ by

$$\begin{aligned}\phi(x_1, x_2, \ldots) &= (x_1 + 1, x_2, x_3, \ldots) \qquad \text{if} \quad x_1 < k_1 \\ &= (0, \ldots, 0, x_p + 1, x_{p+1}, x_{p+2}, \ldots) \\ &\qquad \text{if} \quad x_1 = k_1, \ldots, x_{p-1} = k_{p-1}, \quad x_p < k_p \\ \phi(k_1, k_2, \ldots) &= (0, 0, 0, \ldots).\end{aligned}$$

Show, as in Example 3, that the inverse image of a cylinder set is a cylinder set. Conclude that ϕ is measurable, and that it is measure preserving iff p_{nj} is independent of j; namely, $p_{nj} = 1/(k_n + 1)$.

Doubly Stochastic Operators

6. If $(X, \mathscr{B}, \mu)$ is the unit interval or one of a certain class of "decent" measure spaces, then, for each set function $\psi\colon \mathscr{B} \to \mathscr{B}$ which preserves finite and countable unions and intersections and also preserves complements, there exists a measurable point transformation $\phi\colon X \to X$ such that $\psi(B) = \phi^{-1}(B)$ for all $B \in \mathscr{B}$. Thus Proposition 1.2 may be proved by exhibiting such a ψ.

(a) If χ_A is the characteristic function of the set A, show that for T a doubly stochastic isometry on L_2 and for any $A, B \in \mathscr{B}$

$$\int_X (T\chi_A)(T\chi_B)\, d\mu = \int_X T\chi_{A \cap B}\, d\mu.$$

(b) Show that $0 \leq T\chi_A \leq 1$ and hence that $0 \leq (T\chi_A)^2 \leq T\chi_A$.

(c) Use (a) and (b) to show that $(T\chi_A)^2 = T\chi_A$, and hence that $T\chi_A$ is the characteristic function of some set $\psi(A)$.

(d) Show that $T\chi_{A \cap B} \leq \min\{T\chi_A, T\chi_B\}$, and hence that

$$0 \leq T\chi_{A \cap B} = (T\chi_{A \cap B})^2 \leq (T\chi_A)(T\chi_B).$$

(e) Use (a) and (d) and the relations $\chi_{A \cup B} = \chi_A + \chi_B - \chi_{A \cap B}$, $\chi_{\tilde{A}} - 1 - \chi_A$, to conclude that ψ preserves finite intersections, finite unions, and complements.

(f) From $\mu(A) = (\chi_A, 1)$ deduce that ψ preserves measure and hence also countable unions and intersections.

7. Suppose that T is an operator on $L_1(X, \mathscr{B}, \mu)$ where μ is a finite or σ-finite measure, and suppose T satisfies

(i) $f \leq 0 \Rightarrow Tf \geq 0$,
(ii) $\|Tf\|_1 \leq \|f\|_1$,
(iii) $\|Tf\|_\infty \leq \|f\|_\infty$,

where $f \in L_1$ for (i) and (ii), and $f \in L_1 \cap L_\infty$ for (iii). Suppose further that $g \in L_1 \cap L_\infty$.

(a) Show that $(Tg - c)^+ \leq T(g - c)^+$ for any constant c.

(b) Show $\int_X (Tg - c)\, h(Tg, c)\, d\mu \leq \int_X (g - c)\, h(g, c)\, d\mu$, where

$$\begin{aligned} h(u, v) &= 1 \qquad \text{if} \quad u > v \\ &= 0 \qquad \text{if} \quad u \leq v. \end{aligned}$$

(c) Suppose that $g \geq 0$. Multiply both sides of the above inequality by c^{p-2} and integrate with respect to c from 0 to ∞. Apply the Fubini–Tonelli theorem to obtain

$$\left[\frac{1}{p-1} - \frac{1}{p}\right] \int_X (Tg)^p\, d\mu \leq \left[\frac{1}{p-1} - \frac{1}{p}\right] \int_X g^p\, d\mu,$$

and hence $\|Tg\|_p \leq \|g\|_p$.

(d) From (c) and $|Tg| \leq T|g|$ deduce that $\|T\|_p \leq 1$.

8. **(a)** Let T be a doubly stochastic operator. Suppose that $0 \leq f_n \uparrow f$ a.e. with $f \in L_1$. Show that $Tf_n \uparrow Tf$ by showing that $\int_B Tf\, d\mu = \int_B \lim_{n \to \infty} Tf_n\, d\mu$ for each $B \in \mathscr{B}$.

(b) The preceding is a "monotone convergence theorem" for T. Formulate and prove a "dominated convergence theorem" for T.

Ergodic Theorems

9. Let $(X, \mathscr{B}, \mu)$ be a σ-finite measure space. The statement and proof of Theorem 1.1 remain valid in this context.

(a) Show that Corollary 1.1.1 also remains valid as follows. Let

$f_n^*(x) = \max_{1 \le k \le n}(1/k)T_k f(x)$ and $A_n^*(f; \alpha) = \{x : f_n^*(x) > \alpha\}$. For fixed n and any measurable set C with finite measure, let $h = f - \alpha\chi_C$. Deduce as in the proof of Theorem 1.1 that $\int_{B_n^*(h)} h \, d\mu \ge 0$ so that

$$\int_{B_n^*(h)} f(x) \, \mu(dx) \ge \alpha\mu(C \cap B_n^*(h)).$$

(b) If $\{C_j\}$ is an increasing sequence of measurable sets with finite measure and union X, show that as $j \to \infty$ the sequence $B_n^*(f - \alpha\chi_{C_j})$ decreases to $A_n^*(f; \alpha)$ if $\alpha > 0$ or increases to the same limit if $\alpha < 0$. Conclude that

$$\int_{A_n^*(f; \alpha)} f \, d\mu \ge \alpha\mu(A_n^*(f; \alpha)),$$

and complete the proof of Corollary 1.1.1 by letting $n \to \infty$.

(c) The proof of Theorem 1.2 now goes through as before. In particular, $\int |\bar{f}| \, d\mu \le \int |f| \, d\mu$. Show that equality does not always hold by considering the transformation $\phi(x) = x + 1$ on the reals.

10. **(a)** By an appropriate choice of the function g, show that Birkhoff's theorem follows from the Chacon–Ornstein theorem (Theorem 1.3).

(b) Let $P = (p_{ij})$ be an infinite matrix with $\Sigma_j p_{ij} = 1$ for each i and $p_{ij} \ge 0$ for all i, j. Let Z be the integers, and provide it with a measure structure by letting $\mu(A)$ be the number of elements in A. Define T by $T\{f_i\} = \{g_i\}$, where $g_j = \Sigma_i p_{ij} f_i$. Show that T satisfies the hypotheses of Theorem 1.3. Conclude (Kolmogorov's theorem) that

$$\lim_{n \to \infty} \frac{1}{n} \sum_{k=0}^{n-1} p_{ij}^{(k)} = p_{ij}^*$$

exists, where $p_{ij}^{(k)}$ is the (i, j)-entry in P^k. Also (ratio limit theorem)

$$\lim_{n \to \infty} \sum_{k=0}^{n-1} p_{ij}^{(k)} \Big/ \sum_{k=0}^{n-1} p_{mj}^{(k)}$$

exists.

11. (*Mean ergodic theorem*) Let Φ be an abstract dynamical system and $T = T_\phi$.

(a) Suppose $T^*F = F$. By evaluating $\|TF - F\|_2{}^2 = (TF - F, TF - F)$, show directly that $TF = F$. This gives a simplified proof of Theorem 1.4 for $p = 2$.

(b) If $f \in L_p$, then Theorem 1.4 implies that $(1/n)T_n f \to f^*$ in measure. On the other hand, by Theorem 1.2, $(1/n)T_n f \to \bar{f}$ in measure. Hence $\bar{f} = f^*$ a.e. In particular, $\bar{f} \in L_p$.

(c) Show from (b) that $\int_X \bar{f}\,d\mu = \int_X f\,d\mu$. This can also be proved directly by considering the restriction of ϕ to the invariant set

$$B(\alpha, \beta) = \{x : \alpha < \bar{f}(x) \le \beta\}$$

and applying Corollary 1.1.1 to obtain

$$\alpha\mu(B(\alpha, \beta)) \le \int_{B(\alpha,\beta)} f\,d\mu \le \beta\mu(B(\alpha, \beta))$$

$$\alpha\mu(B(\alpha, \beta)) \le \int_{B(\alpha,\beta)} \bar{f}\,d\mu \le \beta\mu(B(\alpha, \beta)).$$

In particular,

$$-\frac{1}{2^n}\mu\left(B\left(\frac{k}{2^n}, \frac{k+1}{2^n}\right)\right) \le \int_{B(k/2^n,\,(k+1)/2^n)} f\,d\mu - \int_{B(k/2n,\,k+1/2^n)} \bar{f}\,d\mu$$
$$\le \frac{1}{2^n}\mu\left(B\left(\frac{k}{2^n}, \frac{k+1}{2^n}\right)\right).$$

Adding on $k = 0, \pm 1, \pm 2, \ldots$ and then letting $n \to \infty$ gives the desired result.

12. Show that f^* is an invariant function (in Theorem 1.4); that is, show that $Tf^* = f^*$.

Recurrence

13. (a) In Theorem 1.5 show that almost every point of A returns to A infinitely often.

(b) Show that the conclusion of Theorem 1.5 fails for the transformations defined in Exercise 2.

14. (a) An operator T on L_∞ is said to be *conservative* if $f \in L_\infty$, $f \ge 0$, $\sum_{n=0}^{\infty} T^n f(x) < \infty$ a.e. implies that $f = 0$. Show that any doubly stochastic operator on L_∞ of a finite measure space is conservative.

(b) If $\phi: X \to X$ is any measurable transformation on a σ-finite measure space, and if T_ϕ is conservative, then the sequence $A, \phi^{-1}(A), \phi^{-2}(A), \ldots$ can be pairwise disjoint only if $\mu(A) = 0$. Hence ϕ fulfills the conclusion of Theorem 1.5.

Ergodicity and Mixing

15. If $(X, \mathscr{B}, \mu)$ is a finite measure space, and if ϕ is a measure-preserving transformation, show that $\phi^{-1}(B) \subseteq B$ implies $\phi^{-1}(B) = B$. Thus ϕ is ergodic iff $\phi^{-1}(B) \subseteq B \Rightarrow \mu(B) = 0$ or $\mu(\tilde{B}) = 0$. Show that the two definitions are not equivalent in the case of a σ-finite measure space. We adopt the latter as our definition of ergodicity in that case.

16. Let Φ be an abstract dynamical system. Show that the following are equivalent:

(a) Φ is ergodic.

(b) $T_\phi f = f$, $f \in L_p \Rightarrow f$ is a constant.

(c) For all $f \in L_p$,

$$\lim_{n\to\infty} \frac{1}{n} \sum_{k=0}^{n-1} T_\phi^k f(x) = \int_X f\, d\mu \qquad \text{a.e.}$$

(d) For all $f \in L_p$,

$$\lim_{n\to\infty} \left\| \frac{1}{n} \sum_{k=0}^{n-1} T_\phi^k f - \int_X f\, d\mu \right\|_p = 0.$$

(e) For all $f \in L_p$, $g \in L_q$, where $1/p + 1/q = 1$,

$$\lim_{n\to\infty} \frac{1}{n} \sum_{k=0}^{n-1} (f, T_\phi^k g) = (f, 1)(1, g).$$

(f) For all $A, B \in \mathscr{B}$,

$$\lim_{n\to\infty} \frac{1}{n} \sum_{k=0}^{n-1} \mu(A \cap \phi^{-k}(B)) = \mu(A)\, \mu(B).$$

(g) For all $A, B \in \mathscr{B}$ with $\mu(A)\, \mu(B) > 0$,

$$\sum_{n=1}^{\infty} \mu(A \cap \phi^{-n}(B)) > 0.$$

(h) For all $A, B \in \mathscr{B}$ with $\mu(A)\, \mu(B) > 0$,

$$\sum_{n=1}^{\infty} \mu(A \cap \phi^{-n}(B)) = +\infty.$$

17. Let Φ be an abstract dynamical system. Show that the following are equivalent:

(a) Φ is weakly mixing.

(b) For all $f, g \in L_2$ there exists a set J of density zero such that

$$\lim_{n\to\infty,\, n\notin J} (f, T_\phi^n g) = (f, 1)(1, g).$$

(c) For all $f, g \in L_2$

$$\lim_{n\to\infty} \frac{1}{n} \sum_{k=0}^{n-1} |(f, T_\phi^k g)|^2 = |(f, 1)(1, g)|^2.$$

18. Let ϕ_0 be an ergodic measure-preserving transformation on [0, 1]. Let X consist of the two disjoint line segments $X_1 = \{(x, 0) : 0 \leq x \leq 1\}$ and $X_2 = \{(x, 1) : 0 \leq x \leq 1\}$ with linear measure normalized to one. Define ϕ on $X = X_1 \cup X_2$ by $\phi(x, 0) = (x, 1)$ and $\phi(x, 1) = (\phi_0(x), 0)$.

(a) Show that ϕ is an ergodic measure-preserving transformation.

(b) Show that $\mu(X_1 \cap \phi^{-n}(X_2))$ takes on only the values 0 and $\frac{1}{2}$, hence does not converge.

19. (a) If $\mathscr{A}$ is an algebra of subsets of X, and if $\mathscr{B} = \mathscr{B}(\mathscr{A})$ (see Exercise 1), then for each $A, B \in \mathscr{B}$ and each $\varepsilon > 0$ there exist sets $A_0, B_0 \in \mathscr{A}$ such that

$$|\mu(A \cap \phi^{-k}(B)) - \mu(A_0 \cap \phi^{-k}(B_0))| \leq \mu[(A \cap \phi^{-k}(B)) \,\Delta\, (A_0 \cap \phi^{-k}(B_0))]$$
$$\leq \mu[(A \,\Delta\, A_0) \cup (\phi^{-k}(B \,\Delta\, B_0))] < 2\varepsilon$$

for all k.

(b) If $\mathscr{C}$ is a class of subsets of X satisfying condition (30) of Exercise 1, and if $\mathscr{B} = \mathscr{B}(\mathscr{C})$, then Φ is (1) ergodic, (2) weakly mixing, or (3) mixing iff the defining relation is satisfied for all $A, B \in \mathscr{C}$.

Products and Factors

20. Give an example of ergodic systems Φ_1 and Φ_2 such that $\Phi_1 \otimes \Phi_2$ is not ergodic.

21. Show that the union of a finite number of sets of density zero has density zero. Show that this is false for a countable number.

22. If $\mathscr{B}_1 \subseteq \mathscr{B}$, show that $\Phi = (X, \mathscr{B}_1, \mu, \phi)$ is an invertible dynamical system iff $\phi^{-1}(\mathscr{B}_1) = \mathscr{B}_1$.

Inverse Limits

23. Show that the system Φ of Exercise 5 is an inverse limit of the sequence $\Phi_n = (Y_n, \mathscr{A}_n, \nu_n, \phi_n)$, where

$$Y_n = \mathop{\times}_{k=1}^{n} X_k, \qquad \mathscr{A}_n = \mathop{\times}_{k=1}^{n} \mathscr{B}_k, \qquad \nu_n = \mathop{\times}_{k=1}^{n} \mu_k,$$

and

$$\begin{aligned} \phi_n(x_1, \ldots, x_n) &= (x_1 + 1, x_2, \ldots, x_n) \\ &\qquad \text{if} \quad x_1 < k_1 \\ &= (0, \ldots, 0, x_p + 1, x_{p+1}, \ldots, x_n) \\ &\qquad \text{if} \quad x_1 = k_1, \ldots, x_{p-1} = k_{p-1}, \quad x_p < k_p \\ \phi_n(k_1, k_2, \ldots, k_n) &= (0, 0, \ldots, 0). \end{aligned}$$

24. A *Lebesgue system* is an abstract dynamical system $\Phi = (X, \mathscr{B}, \mu, \phi)$ such that there exists a countable class $\mathscr{C} \subseteq \mathscr{B}$ with $\mathscr{B}(\mathscr{C}) = \mathscr{B}$. Show that the inverse limit of a countable number of Lebesgue systems, and hence also the direct product of a countable number of Lebesgue systems, is a Lebesgue system.

25. A *Kolmogorov system* is an invertible dynamical system $\Phi = (X, \mathscr{B}, \mu, \phi)$ for which there is a σ-algebra $\mathscr{B}_0 \subseteq \mathscr{B}$ satisfying

(i) $\phi^{-1}(\mathscr{B}_0) \subseteq \mathscr{B}_0$,
(ii) $\bigcap_{n=1}^{\infty} \phi^{-n}(\mathscr{B}_0) = \{\varnothing, X\}$, and
(iii) $\mathscr{B}(\bigcup_{n=1}^{\infty} \phi^n \mathscr{B}_0) = \mathscr{B}$.

An *exact dynamical system* is a dynamical system $\Phi = (X, \mathscr{B}, \mu, \phi)$ satisfying

(iv) $\bigcap_{n=1}^{\infty} \phi^{-n}(\mathscr{B}) = \{\varnothing, X\}$.

Show that Φ is a Kolmogorov system iff it is the natural extension of an exact system.

26. Show that the two-sided shift on k points is isomorphic to the natural extension of the one-sided shift on k points. Show also that the two-sided shift is Kolmogorov.

27. Prove Propositions 1.9 and 1.10.

Induced Systems

28. Show that Theorem 1.9 and Proposition 1.12 remain valid if ϕ is a recurrent (i.e., one for which the recurrence theorem is valid) measure-preserving transformation of a σ-finite measure space.

29. Prove Theorem 1.10.

30. Construct an example of an ergodic measure-preserving transformation on the reals. Show that any such transformation is conservative, hence recurrent.

31. If Φ_A is induced by $\Phi = (X, \mathscr{B}, \mu, \phi)$ on A, then Φ_A is also induced by $\Phi' = (Y, \mathscr{B}_Y, \mu_Y, \phi_Y)$, where $Y = \bigcup_{n=0}^{\infty} \phi^{-n}(A)$ is a ϕ-invariant subset of X. Moreover, Y is the minimal subset of X for which this is true.

(a) Show that Φ_A is ergodic iff Φ' is ergodic.
(b) Show by example that (a) is false if "ergodic" is replaced by "mixing" or "weak mixing."

Special Flows

32. Let $\Phi = (X, \mathscr{B}, \mu, \phi)$ be a dynamical system, and let $f \geq 0$ be a nonnegative measurable function defined on X. Let Y be the space under the graph of f, that is, $Y = \{(x, y) : x \in X, y \in R, 0 \leq y < f(x)\}$. Y inherits a measure structure as a measurable subset of the product space $X \times R$. Define a family ϕ_t, $0 \le t < \infty$, of transformations of Y as follows:

$$\begin{aligned}\phi_t(x, y) &= (x, y+t), \qquad 0 \le y+t < f(x)\\ &= \left(\phi^n(x), y + t - \sum_{k=0}^{n-1} f(\phi^k(x)),\right.\\ &\qquad \left.\sum_{k=0}^{n-1} f(\phi^k(x)) \le y+t < \sum_{k=0}^{n} f(\phi^k(x))\right).\end{aligned}$$

If t is thought of as time, the point (x, y) moves upward with velocity one until it reaches the "roof" of the space Y, then moves back to the "floor" X and is transformed by ϕ.

(a) Show that ϕ_t is a measure-preserving transformation of Y for each $t \ge 0$.

(b) Show that the transformations ϕ_t form a *flow* in the sense that $\phi_t \phi_s = \phi_{t+s}$ for each $t, s \ge 0$.

(c) Show that the flow is measurable, in the sense that $\phi_{.}(\cdot): Y \times R^+ \to Y$ is a measurable function.

The flow defined in Exercise 31 is called the *special flow* constructed under the function f on the system Φ. Special flows were introduced by Ambrose [5], who showed that all ergodic measurable flows are isomorphic to special flows.

33. In the construction preceding Theorem 1.10, let f be the step function defined on B_0 by setting $f(x) = n+1$ if x is in the range of $\sigma_0 \sigma_1 \cdots \sigma_{n-1}$, but is not in the range of $\sigma_0 \sigma_1 \cdots \sigma_n$ (x lies "under" B_n but not B_{n+1}), and $f(x) = 1$ if x is not in the range of σ_0. If ψ_t is the special flow constructed under f on Φ_0, show that the ϕ of Theorem 1.10 is isomorphic to a factor of ψ_1. (In the construction of the flow ψ_t, look at the subalgebra of sets generated by vertical "columns" between floors.) Does this imply that every ergodic transformation can be embedded in a flow?

34. If $\Phi_1 \prec \Phi_2$, show that any flow ϕ_t constructed on Φ_2 is isomorphic to a flow ψ_t constructed on Φ_1; that is, there exists an invertible measure-preserving transformation σ such that $\sigma\phi_t = \psi_t \sigma$ for all t.

Mixing Transformations

35. **(a)** Let Φ_0 be as in Kakutani's example at the end of Section 6. Show that Φ_0 is isomorphic to the adding machine of Exercise 5 with $k_n = 1$ for each n. Deduce from Exercise 23 and Proposition 1.11 that Φ_0 is ergodic but not weakly mixing.

(b) Set $f_0 \equiv 1$, and, for each $k = 1, 2, 3, \ldots, 2^{n-1}$ and $n = 1, 2, 3, \ldots$, define $f_{k,n}$ on the space of Exercise 5 by

$$f_{k,n}(x_1, x_2, \ldots) = \exp\{2\pi i(2k-1)2^{-(n+1)} \sum_{j=1}^{n} 2^j x_j\}.$$

Show that $f_{k,n}$ is an eigenfunction of T_{ϕ_0} with eigenvalue

$$\lambda_{k,n} = \exp\left\{2\pi i\left(\frac{2k-1}{2^n}\right)\right\}.$$

Note that the range of $f_{k,n}$ is just exactly the 2^nth roots of unity in some order, and that $\lambda_{k,n}$ is also a 2^nth root of unity.

(c) Show that f_0 and the $f_{k,n}$'s constitute a complete orthonormal system in complex $L_2(B_0)$. [*Hint*: Show first that this is true for each $L_2(X_n)$ as defined in Exercise 23.]

(d) For an arbitrary $f \in L_2(B_0)$, expand f in a Fourier series with respect to the basis of (c). Use this expansion, Parseval's relation, and the identity $\lambda_{k,n}^{2^n} = 1$ to deduce

$$\lim_{n\to\infty} \int_{B_0} |f(\phi^{2^n}(x)) - f(x)|^2 \, \mu(dx) = 0.$$

36. Let Φ be as in Kakutani's example. Note that in the terms of Exercise 35 the set A over which B_1 is constructed is the set of $x = \{x_n\}$ for which the smallest n with $x_n = 0$ is odd. (See Fig. 2.) Define u_n on B_0 by $u_1(x) = \chi_A(x) + 1$, and

$$u_n(x) = \sum_{k=0}^{n-1} u_1({\phi_0}^k(x)), \qquad x \in B_0.$$

Note that $u_n(x)$ is n plus the number of "visits" to A in the orbit $x, \phi_0(x), \ldots, \phi_0^{n-1}(x)$. For $n = 4^p$, regardless of what x is, the first $2p$ coordinates $y_1, \ldots, y_{2p}$ take each of the 4^p possible combinations of values exactly once as $y = {\phi_0}^k(x)$ traverses this orbit, y_{p+1} increases by 1, and no other change occurs in y_j, $j > 2p$. For all but one of these combinations, namely $1, 1, \ldots, 1$, $y \in A$ or $y \notin A$ regardless of the values of y_j, $j > 2p$. In the exceptional case, $y \in A$ for exactly $\frac{2}{3}$ of the x values.

(a) Show that $u_n(x)$ for $n = 4^p$ takes only two values, a_p and $a_p + 1$, where

$$a_p = 4^p + 4^p\left(\frac{1}{2} + \frac{1}{2^3} + \cdots + \frac{1}{2^{2p-1}}\right) = \frac{1}{3}(5 \cdot 4^p - 1),$$

and that

$$\mu_0\{x : u_{4^p}(x) = a_p\} = \tfrac{1}{3}, \qquad \mu_0\{x : u_{4^p}(x) = a_p + 1\} = \tfrac{2}{3}.$$

(b) Show that

$$\phi_0{}^n(x) = \phi^{u_n(x)}(x)$$

for each $n = 1, 2, \ldots$ and each $x \in B_0$.

(c) If $f \in L_2(X)$ is such that $f(\phi(x)) = e^{2\pi i\lambda} f(x)$ for all $x \in X$, show that

$$\int_{B_0} |f(\phi^{4^p}(x)) - f(x)|^2 \, \mu_0(dx) = \tfrac{1}{3}|e^{2\pi i\lambda a_p} - 1|^2 + \tfrac{2}{3}|e^{2\pi i\lambda(a_p+1)} - 1|^2.$$

Conclude from Exercise 35(d) that $\lambda = 0$. According to the converse of Theorem 1.6(ii), Φ is weakly mixing.

(d) Use (a) and (b) to show that

$$B_0 \cap \phi^{-a_p}(C) \subseteq \phi_0^{-4^p}(C) \cup \phi_0^{-4^p}(\phi(C))$$

for any measurable set C. In particular, if $C = \{x \in B_0 : x_1 = 0\}$, then the right-hand side reduces to

$$\phi_0^{-4^p}(C) = C.$$

Therefore,

$$\mu((B_0 \sim C) \cap \phi^{-a_p}(C)) = 0$$

for all $p = 1, 2, \ldots$, and ϕ is not strongly mixing.

CHAPTER II

Topological Dynamics

1. CLASSICAL DYNAMICAL SYSTEMS

Topological dynamics may be defined as the study of continuous transformations, or groups of such transformations, defined on a topological space (usually compact), with particular regard to properties of interest in the qualitative theory of differential equations. We shall be concerned in this chapter with the theory of a single continuous transformation of a compact Hausdorff space. Many of the properties of transformation groups (as discussed in [23], for example) may just as well be isolated and studied for a single transformation and its iterates, and we find this study notationally and conceptually much easier to introduce at this level. On the other hand, it should be acknowledged that the classical applications to differential equations and to physics generally involve a continuous group of transformations.

Our considerations on the proper level of generality for this chapter are also guided by our concern for tying in the results obtained here with those of Chapters I and III. For this reason, we shall *not* make the simplifying assumption that our topological space is metrizable. On the other hand, we shall try to point out those situations where a definition or result has a significantly simpler statement in the case of a transformation defined on a metric space.

Definition 2.1 A (*classical*) *dynamical system* is a pair $\Sigma = (X, \sigma)$, where X is a nonempty compact Hausdorff space, and σ is a continuous map of X into itself. Σ is an *invertible* system if σ is invertible.

Note that for an invertible system σ^{-1} is necessarily continuous, and so $\Sigma^{-1} = (X, \sigma^{-1})$ is also a dynamical system.

Although the development in this chapter will be largely independent of the preceding chapter and parallel to it, let us mention a classical result which binds the two theories together. In order to do this, we need first to talk about the operator T_σ induced by σ on the space $C(X)$ of continuous real-valued functions on X.

For $\sigma: X \to X$ a continuous map, we define $T_\sigma: C(X) \to C(X)$ by $T_\sigma f(x) = f(\sigma(x))$. Recall that $C(X)$ is an algebra of functions, and that with the norm

$$\|f\| = \max_{x \in X} |f(x)|,$$

$C(X)$ becomes a complete normed linear space. (See, for example, [16].) It follows easily that T_σ has the following properties:

1. $f \geq 0 \Rightarrow T_\sigma f \geq 0$;
2. $\|T_\sigma f\| \leq \|f\|$;
3. $T_\sigma 1 = 1$.

(Inequalities and equalities of functions are assumed to hold pointwise.) If σ is epic ($\sigma(X) = X$), then property 2 may be replaced by the stronger

2′. $\|T_\sigma f\| = \|f\|$.

If σ is invertible, then T_σ is invertible with $T_\sigma^{-1} = T_{\sigma^{-1}}$.

The dual of the space $C(X)$ as a normed linear space is the space $M(X)$ of all finite (signed) Borel measures on X. This duality is expressed by

$$(f, \mu) = \mu(f) = \int_X f(x)\, \mu(dx).$$

The *adjoint* $T^*: M(X) \to M(X)$ is defined by $(f, T^*\mu) = (Tf, \mu)$, where T is any continuous linear operator on $C(X)$. The topology on $M(X)$ is given by the total variation norm, $\|\mu\| = |\mu|(X)$, and T^* is continuous for this topology. Indeed, $\|T^*\| = \|T\|$. Properties 1–3 for T_σ are equivalent to the following properties for T_σ^*:

1*. $\mu \geq 0 \Rightarrow T_\sigma^*\mu \geq 0$;
2*. $\|T_\sigma^*\mu\| \leq \|\mu\|$;
3*. $(T_\sigma^*\mu)(X) = \mu(X)$.

If σ is monic, then for each function g of norm one there is an f of norm one with $T_\sigma f(x) = f(\sigma(x)) = g(x)$ $(x \in X)$. It follows easily that, in this case, we can replace 2* with the stronger

2'*. $\|T_\sigma{}^*\mu\| = \|\mu\|$.

In particular, if Σ is invertible, then both T_σ and $T_\sigma{}^*$ are invertible isometries, and $T_\sigma{}^{*-1} = T_{\sigma^{-1}}{}^* = T_\sigma{}^{-1*}$.

Now let

$$K = \{\mu \in M(X) : \mu \geq 0, \|\mu\| = \mu(X) = 1\}.$$

The set K is nonempty, convex, and weak*-compact. (See [16], Corollary V.4.3.) Moreover, $T_\sigma{}^*(K) \subseteq K$. According to the Markov–Kakutani fixed point theorem ([16], p. 456), $T_\sigma{}^*$ has a fixed point $\mu \in K$, that is, $T_\sigma{}^*\mu = \mu$. But a fixed point for $T_\sigma{}^*$ is nothing more nor less than an invariant Borel measure for σ, since

$$\int_X T_\sigma f(x)\,\mu(dx) = (T_\sigma f, \mu) = (f, T_\sigma{}^*\mu) = (f, \mu) = \int_X f(x)\,\mu(dx) \tag{1}$$

for all $f \in C(X)$, and validity of (1) on $C(X)$ is equivalent to

$$\mu(\sigma^{-1}(B)) = \mu(B) \tag{2}$$

for all Borel sets B. Thus we have proved the following theorem:

Theorem 2.1 *Let $\Sigma = (X, \sigma)$ be a classical dynamical system. Then there exists a normalized (total measure one), positive measure μ on the class $\mathscr{B}$ of Borel sets of X such that σ preserves the measure μ. That is, $\bar{\Sigma} = (X, \mathscr{B}, \mu, \sigma)$ is an abstract dynamical system.*

Remark We have not required that σ be epic. On the other hand, it is immediate that a measure-preserving transformation must be essentially onto. This shows that the measure μ in the theorem may be degenerate. For example, if $\sigma(x) = x_0$ is a constant map, then μ is concentrated at the single point x_0. We shall see later (Exercise 7) that certain systems must have invariant measures μ whose support, that is, the smallest closed subset of X whose complement has μ-measure zero, is all of X.

Example 1 *(Symbolic dynamical systems)* Let

$$X_n = \{0, 1, \ldots, k-1\}$$

be a finite set of k points with the discrete topology. Form the product

$$X = \mathop{\times}_{n=-\infty}^{\infty} X_n$$

with the product topology. Thus X is a compact, totally disconnected, Hausdorff space. In case $k = 2$, it is homeomorphic to the Cantor "middle-thirds" set.

Define $\sigma: X \to X$ by $\sigma(x) = y$, where $y_n = x_{n+1}$ for all n. Since the "cylinder sets"

$$C = \{x \in X : (x_{n_1}, \ldots, x_{n_l}) \in A\} = \bigcup_{(s_1, \ldots, s_l) \in A} \bigcap_{j=1}^{l} \{x : x_{n_j} = s_j\}$$

constitute a base for the topology of X, and since

$$\sigma^{-1}(C) = \bigcup_{(s_1, \ldots, s_l) \in A} \bigcap_{j=1}^{l} \{x : x_{n_j+1} = s_j\}$$

is also a cylinder set, σ is continuous. Clearly σ is invertible, and $\sigma^{-1}(x) = z$, where $z_n = x_{n-1}$. Thus $\Sigma = (X, \sigma)$ is an invertible dynamical system. It is called a *shift dynamical system* or *symbolic dynamical system* (on the symbols or "alphabet" $\{0, 1, \ldots, k-1\}$).

The measure μ which assigns to the set C above the mass

$$\mu(C) = \sum_{(s_1, \ldots, s_l) \in A} \prod_{j=1}^{l} p_{s_j},$$

where $p_j = 1/k$ for $j = 0, 1, \ldots, k-1$, is invariant for σ. However, as seen in the previous chapter, there are many other invariant measures for σ.

2. MINIMAL AND STRICTLY ERGODIC SYSTEMS

An important and central notion in the study of dynamical systems is that of minimality. The essential idea of minimality is that everything worth knowing about the system can be determined from the present and future, or from the past, present, and future, situation of a single point under the action of σ.

Let $\Sigma = (X, \sigma)$ be a dynamical system. By the *positive orbit* of a point $x \in X$ is meant the set

$$O_\sigma^+(x) = O^+(x) = \{\sigma^n(x) : n = 0, 1, 2, \ldots\} = \bigcup_{n=0}^{\infty} \sigma^n(\{x\}).$$

By the *orbit* of x is meant the set $O_\sigma(x) = O(x) = \bigcup_{n=-\infty}^{\infty} \sigma^n(\{x\})$. We denote the closure of $O^+(x)$ by $\bar{O}^+(x)$ and the closure of $O(x)$ by $\bar{O}(x)$, and refer to these sets as the *positive orbit closure* and the *orbit closure* of x, respectively.

Definition 2.2 The dynamical system $\Sigma = (X, \sigma)$ is *minimal* if $\sigma(A) \subseteq A$, A closed, implies either $A = X$ or $A = \varnothing$.

Definition 2.3 The dynamical system $\Sigma = (X, \sigma)$ is *(positively) recurrent* if $x \in \bar{O}^+(\sigma(x))$ for each $x \in X$, that is, if for each open set $U \subseteq X$ and each $x \in U$ there exists a positive integer n with $\sigma^n(x) \in U$.

Remark If $x \in \bar{O}^+(\sigma(x))$, then either $x = \sigma(x)$ or $x \in \bar{O}^+(\sigma^2(x))$. But in the former case, $x = \sigma^n(x)$ for all n, and so $x \in \bar{O}^+(\sigma^2(x))$. By induction, $x \in \bar{O}^+(\sigma^n(x))$ for all $n \geq 1$. It follows that $\sigma^n(x)$ returns to each neighborhood of x infinitely often.

Proposition 2.1 *The system $\Sigma = (X, \sigma)$ is minimal iff $\bar{O}^+(x) = X$ for each $x \in X$. In particular, if Σ is minimal, then it is recurrent.*

Proof For a given $x \in X$ the set $O^+(x)$ is invariant, that is, $\sigma(O^+(x)) \subseteq O^+(x)$. It follows that its closure is also invariant:

$$\sigma(\bar{O}^+(x)) \subseteq \sigma(O^+(x))^- \subseteq \bar{O}^+(x).$$

If Σ is minimal, since $x \in \bar{O}^+(x) \neq \varnothing$, it follows that $\bar{O}^+(x) = X$.

Conversely, suppose that Σ is not minimal. Then there is a nonempty, closed subset $A \subseteq X$ with $A \neq X$ and $\sigma(A) \subseteq A$. If $x \in A$, then $O^+(x)$ and hence $\bar{O}^+(x)$ are contained in A. Thus $\bar{O}^+(x) \neq X$.

If Σ is minimal and $x \in X$, then $\bar{O}^+(\sigma(x)) = X$, and so $x \in \bar{O}^+(\sigma(x))$. That is, Σ is recurrent. ▮

Proposition 2.2 *If Σ is recurrent, then σ is epic. If, moreover, $\sigma(A) \subseteq A$ for some closed subset A of X, then $\sigma(A) = A$. In particular, $\bar{O}(x) = \bar{O}^+(x)$ for each $x \in X$.*

Proof Suppose that σ is not epic, and let $x \in U = X \sim \sigma(X)$. The set U is open and $\sigma^n(x) \notin U$ for any positive n. Thus Σ is not recurrent.

Now suppose Σ is recurrent and $\sigma(A) \subseteq A$ for some closed subset A of X. Let $x \in A$ and choose $y \in \sigma^{-1}(\{x\})$. Then $y \in \bar{O}^+(\sigma(y)) = \bar{O}^+(x) \subseteq A$. Hence $x = \sigma(y) \in \sigma(A)$. Thus $\sigma(A) = A$. Likewise, for each $x \in X$, $O(x)$ and hence $\bar{O}(x) \subseteq \bar{O}^+(x)$. ▮

We are now in a position to summarize our information about minimal dynamical systems.

Theorem 2.2 *Let $\Sigma = (X, \sigma)$ be a dynamical system. Then there exists a nonempty closed subset $X_0 \subseteq X$ with $\sigma(X_0) \subseteq X_0$ and $\Sigma_0 = (X_0, \sigma)$ minimal. Moreover, any such X_0 must satisfy $\sigma(X_0) = X_0$.*

If $\Sigma = (X, \sigma)$ and X_0 is a closed subset of X, then the following are equivalent:

1. *$\sigma(X_0) \subseteq X_0$ and $\Sigma_0 = (X_0, \sigma)$ is minimal;*
2. *$\bar{O}^+(x) = X_0$ for each $x \in X_0$;*
3. *$\bar{O}(x) = X_0$ for each $x \in X_0$;*
4. *$\sigma(X_0) = X_0$ and X_0 has no closed, nonempty, proper subset A satisfying $\sigma(A) = A$.*

Note If (X_0, σ) is minimal, we shall say that X_0 is a *minimal set* for σ.

Proof Let $\mathscr{C}$ be the class of nonempty closed subsets A of X with $\sigma(A) \subseteq A$. $\mathscr{C}$ is partially ordered by set inclusion. If $\mathscr{C}_0$ is any totally ordered subset of $\mathscr{C}$, then $Y_0 = \cap \mathscr{C}_0$ is a lower bound for $\mathscr{C}_0$. Moreover, Y_0 is closed, and $\sigma(Y_0) \subseteq Y_0$. By compactness of X and the fact that $\mathscr{C}_0$ is totally ordered, $Y_0 \neq \varnothing$. Thus $Y_0 \in \mathscr{C}$. It follows by Zorn's lemma that $\mathscr{C}$ has a minimal element X_0. Thus $\Sigma_0 = (X_0, \sigma)$ is minimal. Moreover, since $\sigma(X_0) \in \mathscr{C}$, it follows that $X_0 = \sigma(X_0)$.

Now suppose that X_0 is a closed subset of X with $\sigma(X_0) \subseteq X_0$. It follows from Proposition 2.1 that statement 1 is equivalent to statement 2, and by a similar argument that statement 3 is equivalent to statement 4. If statement 1 holds, then $\sigma(X_0) = X_0$, and so statement 1 implies statement 4. Let us show, conversely, that statement 4 implies statement 1.

Suppose that $\sigma(X_0) \subseteq X_0$ and Σ_0 is not minimal. By the first part of the theorem, there exists a closed subset $X_1 \subseteq X_0$ with $\Sigma_1 = (X_1, \sigma)$ minimal. Then $X_1 \neq \varnothing$, and since Σ_0 is not minimal, $X_1 \neq X_0$. Again from the first part of the theorem $\sigma(X_1) = X_1$. This shows that statement 4 is not satisfied. ▮

Corollary 2.2.1 *If $A \subseteq X$, $\sigma(A) = A$, A closed, implies that $A = \varnothing$ or X, or equivalently, if $\bar{O}(x) = X$ for each $x \in X$, then Σ is minimal.*

Proof The last part of the proof did not use the fact that $\sigma(X_0) = X_0$. ▮

Corollary 2.2.2 *If Σ is invertible and minimal, then Σ^{-1} is minimal.*

Example 2 Let $X = [0, 1]$ and $\sigma(x) = x^2$. Then $\sigma^n(x) = x^{2^n}$. Σ is invertible, but not minimal or recurrent. In fact, the only recurrent points are the fixed points 0 and 1, and the only minimal sets are $\{0\}$ and $\{1\}$. Sets of the form $A = [0, a]$, $0 < a < 1$, satisfy $\sigma(A) \subseteq A$, but not $\sigma(A) = A$. Sets of the form $\bar{O}^+(a) = \{0, a, a^2, a^4, \ldots\}$ are closed and invariant, but not minimal or recurrent. Sets of the form $A = \bar{O}(a) = \{0, 1\} \cup \{a^{2^n} : n \in Z\}$ satisfy $\sigma(A) = A$ and $A = \bar{O}(x)$ for all $x \in A$ except 0 and 1.

Example 3 Let $X = [0, 1]$ and let $\sigma(x)$ be the fractional part of $x + a$. If a is rational, then each point has finite orbit, and each orbit is a minimal set. If a is irrational, then $O^+(x)$ is dense in X for each $x \in X$, and so Σ is minimal.

Example 4 Let Σ be the symbolic dynamical system (two-sided shift) on two points. Choose $x \in X$ by setting x_n equal mod 2 to the sum of the binary digits in the binary expansion of n for $n \geq 0$, that is,

$$x_n = \sum_{j=0}^{k} a_j \pmod 2 \qquad \text{where} \quad n = \sum_{j=0}^{k} a_j 2^j,$$

and setting $x_{-n} = x_{n-1}$. Then (Exercise 5) the closed invariant set $X_0 = \bar{O}(x)$ is minimal.

The definition of a minimal dynamical system bears a strong resemblance to that of an ergodic abstract dynamical system (Definition 1.4). In cases where both apply it is in fact a stronger condition (see Exercise 9). For ergodic systems Φ and integrable functions f, we have seen that the "time averages" $f_n(x) = (1/n)\sum_{k=0}^{n-1} f(\phi^k(x))$ converge μ-a.e. to a constant, namely, the "phase average" $\mu(f) = \int_X f(x)\,\mu(dx)$ with respect to the invariant measure μ. Since topological dynamics is concerned with relations holding at each point of X rather than almost everywhere, we might expect that this convergence would hold for each $x \in X$ in case $\Phi = (X, \phi)$ is a minimal classical dynamical system and $f \in C(X)$. However, this is known to be false ([49], p. 134).

The ergodic theorem and its corollary, Proposition 1.4, give us a clue as to the proper conditions to ensure that $f_n(x)$ converges everywhere to a constant. For if μ_1 and μ_2 are each ergodic invariant measures, then there will be values of x for which $f_n(x) \to \mu_1(f)$ and other values for which $f_n(x) \to \mu_2(f)$. Moreover, if $f_n(x) \to L_x(f)$ for a given $x \in X$ and for all $f \in C(X)$, then L_x determines a unique σ-invariant, but not necessarily ergodic, Borel measure μ_x with $\mu_x(f) = L_x(f)$. If f_n is to converge to a constant, we must have μ_x independent of x. In the following, we follow Furstenberg [21], except that our definitions of unique ergodicity and strict ergodicity follow [49], to which the reader is referred for an extremely lucid presentation of the underlying ideas.

Definition 2.4 A dynamical system $\Sigma = (X, \sigma)$ is *uniquely ergodic* if there is exactly one σ-invariant, normalized Borel measure μ on X. Σ is *strictly ergodic* if it is uniquely ergodic and minimal.

A point $x \in X$ is said to be *generic* for μ if $f_n(x) \to \mu(f)$ for each $f \in C(X)$.

Theorem 2.3 *Let $\Sigma = (X, \sigma)$ be a dynamical system. Then the following are equivalent*:

1. *Σ is uniquely ergodic with invariant measure μ;*
2. *$f_n(x)$ converges to $\mu(f)$ uniformly on X, for each $f \in C(X)$;*
3. *every point of X is generic for μ.*

Notice that the equivalence of statements 1 and 3 answers the question raised above about pointwise convergence of $f_n(x)$, and that the uniform convergence in statement 2 is a bonus.

Proof It is clear that statement 2 implies statement 3. We shall show that statement 1 implies statement 2 and statement 3 implies statement 1. Suppose that statement 1 is true. The uniqueness of the invariant measure μ means that the set of measures $\nu \in M(X)$ that vanish on the subspace $\mathscr{X}_1 = \{g - T_\sigma g : g \in C(X)\}$ of $C(X)$ is one-dimensional. It follows that the closure in $C(X)$ of $\mathscr{X}_1$ coincides with the null space $\mathscr{X}_\mu = \{f - \mu(f) : f \in C(X)\}$ of μ. (The annihilator of the annihilator of $\mathscr{X}_1$ is the closure of $\mathscr{X}_1$.) Thus, given $f \in C(X)$ and $\varepsilon > 0$, there exists $g \in C(X)$ with $\|f - \mu(f) - (g - T_\sigma g)\| < \varepsilon$. It follows that

$$\left\| f_n - \mu(f) - \frac{1}{n}(g - T_\sigma{}^n g) \right\| \leq \frac{1}{n} \sum_{k=0}^{n-1} \|T_\sigma{}^n(f - \mu(f) - (g - T_\sigma g))\| < \varepsilon$$

for each n. Since $(1/n)(g - T_\sigma{}^n g)$ converges uniformly to zero, it follows that $\|f_n - \mu(f)\| \to 0$ as $n \to \infty$.

Now suppose that statement 3 holds, and that ν is any normalized, invariant Borel measure on X. Since the functions $f_n(x)$ are uniformly bounded by $\|f\|$ and $f_n(x) \to \mu(f)$ for each $x \in X$, it follows by the bounded convergence theorem that

$$\mu(f) = \lim_{n \to \infty} \int_X f_n(x)\, \nu(dx) = \nu(f)$$

for each $f \in C(X)$. It follows that $\nu = \mu$. ∎

Example The system of Example 2 is not uniquely ergodic since it has two fixed points. However, a close relative of it is. Let $X = K = \{z : |z| = 1\}$ be the unit circle in the complex plane, and define σ by $\sigma(e^{2\pi i x}) = \exp(2\pi i x^2)$. Then Σ is an invertible dynamical system. If μ is any invariant measure, then the μ-measure of the arc from 1 counterclockwise to $e^{2\pi i a}$ is the same as that of the arc from 1 to $\exp(2\pi i a^2)$ for each a, $0 < a < 1$. Hence the arc from $\exp(2\pi i a^2)$ to $e^{2\pi i a}$ has measure zero. It follows that μ is concentrated at the single point $z = 1$, and that Σ is uniquely ergodic. On the other hand, Σ is not minimal, hence not strictly ergodic. An example of a strictly ergodic system is given in Exercise 1.

Remark A uniquely ergodic system is strictly ergodic iff the support of the unique invariant measure μ is all of X, that is, iff $\mu(U) > 0$ for each open set $U \subseteq X$. (See Exercise 7.)

3. EQUICONTINUOUS AND DISTAL SYSTEMS

In this section we shall assume that $\Sigma = (X, \sigma)$ is an invertible dynamical system. It turns out that this is only an apparent restriction for distal systems, as we shall see in Section 6.

Recall that a compact Hausdorff space X can always have its topology described in terms of a uniformity on X. (See, for example, [55].) In particular, if X is metrizable with metric d, then the pair (x, y) belongs to the index $\alpha_\varepsilon \subseteq X \times X$ provided that $d(x, y) < \varepsilon$. A collection $\{f_j : j \in J\}$ of functions from one uniform space X to another uniform space Y is *equicontinuous* if for each index β on Y there exists an index α on X such that $(x, y) \in \alpha$ implies $(f_j(x), f_j(y)) \in \beta$ for all $j \in J$.

Definition 2.5 A dynamical system $\Sigma = (X, \sigma)$ is *equicontinuous* if the collection $\{\sigma^n : n \in Z\}$ of transformations of X is equicontinuous.

Examples The shift dynamical systems of Example 1 are not equicontinuous. A suitable metric on X is given by

$$d(x, y) = (1 + \min\{|n| : x_n \neq y_n\})^{-1}.$$

With this metric, there exist points x and y (differing only in the nth component) with $d(x, y) = 1/(1 + n)$ arbitrarily small but with $d(\sigma^{-n}(x), \sigma^{-n}(y)) = 1$.

Example 3 is not equicontinuous as it stands, but it becomes so if we modify the topological space X by identifying 0 and 1, or, equivalently, by defining σ on $K = \{z : |z| = 1\}$ by $\sigma(z) = e^{2\pi i a}z$, for then $d(\sigma^n(z), \sigma^n(w)) = |e^{2\pi i n a}(z - w)| = d(z, w)$.

Proposition 2.3 *A minimal equicontinuous dynamical system is strictly ergodic.*

Proof Let $f \in C(X)$. Equicontinuity of the transformations σ^n implies equicontinuity of the family of real-valued functions $f(\sigma^n(\cdot))$ $(n \in Z)$ and hence of the functions f_n $(n \in Z)$. It follows from the Arzelà–Ascoli theorem [16, p. 266] that some subsequence f_{n_k} converges uniformly on X. Let g be the uniform limit of this sequence. Then g is continuous, and $T_\sigma g = g$. Since Σ is minimal, g must be a constant (see Exercise 9). Let us indicate this constant, which depends only on the choice of f, by $\lambda(f)$. Now if μ is any invariant normalized Borel measure on X, then $\mu(f_n) = \int_X f_n \, d\mu$ converges to $\mu(g) = \lambda(f)$. But $\mu(f_n) = \mu(f)$. Thus $\mu(f) = \lambda(f)$ for each $f \in C(X)$. It follows that $\mu = \lambda$. ∎

In general, of course an equicontinuous system need not be minimal and so need not be uniquely ergodic. However, we shall see in Section 6 that if $\Sigma = (X, \sigma)$ is equicontinuous, then X is a union of disjoint invariant sets on which σ is minimal.

A closely related notion to that of an equicontinuous system is that of a distal system. There are several equivalent ways of defining distal systems (Exercise 10), and we choose one that is most simply stated in terms of the product topology on the product space $X \times X$.

Definition 2.6 A dynamical system $\Sigma = (X, \sigma)$ is *distal* if, for each pair $x, y \in X$ with $x \neq y$, the closure of the set $\{(\sigma^n(x), \sigma^n(y)) : n \in Z\}$ is disjoint from the diagonal $\Delta = \{(x, x) : x \in X\}$ in $X \times X$.

Proposition 2.4 *If Σ is equicontinuous, then it is distal.*

Proof Suppose $x \neq y$. Then there exists an index β on X with $(x, y) \notin \beta$. By equicontinuity there exists an index α such that $(u, v) \in \alpha$ implies $(\sigma^k u, \sigma^k v) \in \beta$ for all $k \in Z$. It follows that $(\sigma^n x, \sigma^n y) \notin \alpha$ for any $n \in Z$. Otherwise, we could let $u = \sigma^n x$, $v = \sigma^n y$, $k = -n$, and reach a contradiction. Thus $\{(\sigma^n(x), \sigma^n(y) : n \in Z\}$ is disjoint from α. Since $\Delta \subseteq \alpha$ and α is open in the product topology, it follows that Σ is distal. ▮

Remark For a metrizable space X with metric d, the system $\Sigma = (X, \sigma)$ is distal iff for each pair $x, y \in X$ with $x \neq y$, there exists an $\varepsilon > 0$ with $d(\sigma^n(x), \sigma^n(y)) \geq \varepsilon$ for all n. In general (Exercise 10), Σ is distal iff $\sigma^{n_k} x \to z$, $\sigma^{n_k} y \to z$ for some generalized sequence (net) of integers n_k implies that $x = y$.

Examples The shift dynamical (symbolic) systems of Example 1 are not distal. In fact, one can easily find points $x, y \in X$ with $x \neq y$, $\sigma(x) = x$, $\sigma^n(y) \to x$ as $n \to \infty$.

The modification of Example 3 discussed earlier is equicontinuous, hence distal. Let us describe a similar example which is distal but not equicontinuous.

Example 5 Let $X = K \times K$, where $K = \{z : |z| = 1\}$ is the unit circle in the complex plane. Define σ on X by $\sigma(z, w) = (e^{2\pi i a} z, zw)$, where a is an irrational number between 0 and 1. This is the simplest of the so-called skew product transformations introduced by H. Anzai [32, p. 60] and extensively studied both as classical and as abstract dynamical systems. It is not hard to show that the orbit of each point in X is dense and hence that $\Sigma = (X, \sigma)$ is minimal. Let us show that it is distal. We take as metric the product metric on $K \times K \subseteq E_4$.

The transformation σ carries all points (z, w) having the same z-coordinate onto points again having the same z-coordinate and changes the w-coordinate by the same multiplicative factor. Thus if $z_1 \neq z_2$, then

$$d(\sigma^n(z_1, w_1), \sigma^n(z_2, w_2)) \geq d(z_1, z_2) \qquad \text{for all} \quad n.$$

On the other hand, if $z_1 = z_2$, then

$$d(\sigma^n(z_1, w_1), \sigma^n(z_2, w_2)) = d(w_1, w_2).$$

In either case, for $x \neq y$ there exists an $\varepsilon > 0$ with $d(\sigma^n(x), \sigma^n(y)) > \varepsilon$ for all n.

Next we show that Σ is not equicontinuous. Notice that for positive n we have

$$\sigma^n(z, w) = (e^{2\pi ina}z, e^{\pi in(n-1)a}z^n w).$$

Taking $z_n = e^{\pi i/n}z$, we see that $(z_n, w) \to (z, w)$ as $n \to \infty$. Thus $d((z_n, w), (z, w))$ can be made arbitrarily small. On the other hand,

$$d(\sigma^n(z_n, w), \sigma^n(z, w)) \geq d(-w, w) = 2.$$

Taking $\varepsilon = 1$, we see that Σ is not equicontinuous.

4. SUMS AND PRODUCTS OF DYNAMICAL SYSTEMS

We begin now the study of methods of constructing new dynamical systems from given ones. In addition to the products and factors, defined as they were in the previous chapter for abstract systems, we shall consider the categorically dual constructions of sums and subsystems.

Definition 2.7 Let $\Sigma_j = (X_j, \sigma_j)$ be a classical dynamical system for each $j \in J$. We define the *direct product* $\Sigma = \bigotimes_{j \in J} \Sigma_j$ by taking for the space X the product $\times_{j \in J} X_j$ with the product topology, and defining $\sigma = \bigotimes_{j \in J} \sigma_j$ by

$$\sigma(x) = y, \qquad \text{where} \quad y_j = \sigma_j(x_j).$$

We also make use of the customary modifications of this notation, such as $\Sigma_1 \otimes \Sigma_2$, $\Sigma_1 \otimes \cdots \otimes \Sigma_n$, and $\bigotimes_{n=1}^{\infty} \Sigma_n$.

Definition 2.8 Let $\Sigma_j = (X_j, \sigma_j)$ be a classical dynamical system for each $j \in J$, where either (i) J is finite, or (ii) $X = \bigcup_{j \in J} X_j$ is compact Hausdorff, the X_j are pairwise disjoint and have the relative topology as subsets of X. In case (i), define $X = \bigoplus_{j \in J} X_j$ to be the disjoint union of copies of the X_j with the direct sum topology (a set is open if its intersection with each X_j is open), and define $\sigma(x) = \sigma_j(x)$ for $x \in X_j$. In case (ii), assume that $\sigma\colon X \to X$ is continuous and satisfies $\sigma(x) = \sigma_j(x)$ for $x \in X_j$. In either case, we define the *direct sum* $\Sigma = \bigoplus_{j \in J} \Sigma_j$ to be $\Sigma = (X, \sigma)$. Again, we shall make use of the customary modifications. If Σ is a direct sum of minimal systems, we say Σ is *semisimple*.

Remark It is possible to give categorical definitions as follows: If $\Sigma = \bigotimes_{j \in J} \Sigma_j$, then for each $j \in J$ there is a commutative diagram

$$\begin{array}{ccc} X & \xrightarrow{\ \sigma\ } & X \\ {\scriptstyle\psi_j}\downarrow & & \downarrow{\scriptstyle\psi_j} \\ X_j & \xrightarrow{\ \sigma_j\ } & X_j \end{array} \tag{3}$$

where the ψ_j are epimorphisms, and if $\Omega = (Y, \omega)$ completes a similar diagram:

$$\begin{array}{ccc} Y & \xrightarrow{\ \omega\ } & Y \\ {\scriptstyle\rho_j}\downarrow & & \downarrow{\scriptstyle\rho_j} \\ X_j & \xrightarrow{\ \sigma_j\ } & X_j \end{array} \tag{4}$$

with the ρ_j epimorphisms for each $j \in J$, then there is an epimorphism $\phi\colon Y \to X$ such that the diagrams

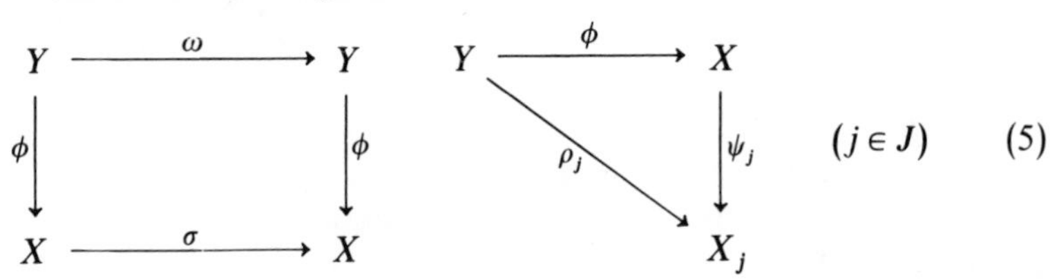

commute. The ψ_j, of course, are the projections $\psi_j(x) = x_j$, and ϕ is defined by

$$\phi(y) = x, \qquad \text{where} \quad x_j = \rho_j(y).$$

Finite direct sums are defined in exactly the same way, with all of the arrows reversed and epimorphism replaced by monomorphism throughout. In this case, the ψ_j are the injections $\psi_j(x) = x$, and ϕ is the monomorphism defined by

$$\phi(x) = \rho_j(x) \qquad (x \in X_j).$$

Infinite direct sums are, in general, not categorical. Uniqueness fails, since the map ϕ defined above may not be continuous.

We may now restate Definition 2.6 as follows.

Definition 2.6′ The dynamical system $\Sigma = (X, \sigma)$ is *distal* provided that the orbit closure of each point for the system $\Sigma \otimes \Sigma$ is either contained in or disjoint from the diagonal Δ.

Example 6 As an example of a direct sum consisting of an infinite number of summands, consider the transformation σ defined on the torus $K \times K$ by $\sigma(z, w) = (e^{2\pi i a}z, e^{4\pi i a}w)$, where a is an irrational number between 0 and 1. This transformation is equicontinuous, but not minimal. In fact, each of the pairwise disjoint curves of the form $w = \lambda z^2$, where λ is a given point in K, is a minimal set for σ. It follows that the systems $\Sigma_\lambda = (X_\lambda, \sigma_\lambda)$ $(\lambda \in K)$, where X_λ is the curve described above and σ_λ is the restriction of σ to X_λ, satisfy the conditions of Definition 2.8.

Each of the X_λ in this example is homeomorphic to the compact group K, and σ_λ is carried by this homeomorphism into a rotation through the angle $2\pi a$. This situation is, as we shall see later, typical of equicontinuous systems.

Diagrams (3)–(5) and their duals for sums suggest two more definitions.

Definition 2.9 We shall say that the dynamical system $\Sigma_1 = (X_1, \sigma_1)$ is a *factor* of the system $\Sigma = (X, \sigma)$ if there exists a continuous epimorphism $\psi: X \to X_1$ such that the diagram

$$\begin{array}{ccc} X & \xrightarrow{\sigma} & X \\ \downarrow{\scriptstyle\psi} & & \downarrow{\scriptstyle\psi} \\ X_1 & \xrightarrow{\sigma_1} & X_1 \end{array} \qquad (6)$$

commutes. In this case, we write $\Sigma_1 | \Sigma$ and $\psi: \Sigma \to \Sigma_1$ or $\Sigma \xrightarrow{\psi} \Sigma_1$. The map ψ is called a *homomorphism* (or epimorphism) of Σ onto Σ_1. If ψ is invertible, it is called an *isomorphism*, and we say Σ_1 and Σ are *isomorphic* dynamical systems. If $\Sigma = \Sigma_1 \otimes \Sigma_2$, we say Σ_1 is a *direct factor* of Σ.

Definition 2.10 We shall say that the dynamical system $\Sigma_1 = (X_1, \sigma_1)$ is a *subsystem* of the system $\Sigma = (X, \sigma)$ if there exists a continuous monomorphism $\psi: X_1 \to X$ such that the diagram

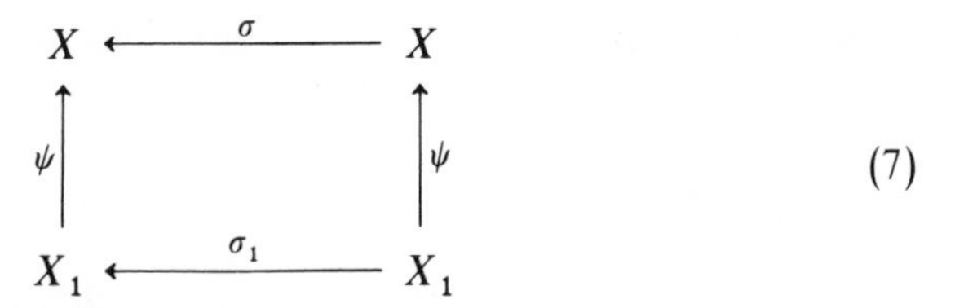

(7)

commutes. The map ψ is called an *injection* (embedding, monomorphism) of Σ_1 into Σ.

To complete the analogy between Definitions 2.9 and 2.10, we should call the system Σ_1 in the latter case a summand of Σ. However, the given terminology is well established and agrees with terminology in other categories, for example, subgroup, subspace, etc. We shall, however, refer to Σ_1 as a *direct summand* in case $\Sigma = \Sigma_1 \oplus \Sigma_2$.

Proposition 2.5 *A dynamical system is minimal iff it has no proper subsystems. A factor of a minimal (equicontinuous) system is minimal (equicontinuous).*

The proof is left as an exercise. Note that nontrivial sums are never minimal. Nor is $\Sigma \otimes \Sigma$ ever minimal, since the diagonal is a closed invariant set. Contrast this with the situation for abstract dynamical systems Φ, where $\Phi \otimes \Phi$ is ergodic when Φ is weakly mixing.

Proposition 2.6 *Products, finite sums, and subsystems of equicontinuous (distal) dynamical systems are equicontinuous (distal). Arbitrary sums of distal systems are distal.*

Proof Consider first the product $\Sigma = \bigotimes_{j \in J} \Sigma_j$. Suppose that each Σ_j is equicontinuous. A basis for the uniformity on X is obtained by taking products $\beta = \mathsf{X}_{j \in J} \beta_j$, where β_j is an index on X_j for all $j \in J$, and all but finitely many $\beta_j = X_j \times X_j$. For the finite number of exceptions, choose α_j so that $(\xi, \eta) \in \alpha_j$ implies $(\sigma_j{}^n(\xi), \sigma_j{}^n(\eta)) \in \beta_j$ for all n. Let $\alpha_j = X_j \times X_j$ otherwise, and set $\alpha = \mathsf{X}_{j \in J} \alpha_j$. Then $(x, y) \in \alpha$ implies $(\sigma^n(x), \sigma^n(y)) \in \beta$ for all n, and Σ is equicontinuous.

Suppose each Σ_j is distal and that there exists a generalized sequence (net) of integers n_k and points $x, y, z \in X$ such that $\sigma^{n_k}(x) \to z$ and $\sigma^{n_k}(y) \to z$. Then for each $j \in J$ it follows that $\sigma_j^{n_k}(x_j) \to z_j$, $\sigma_j^{n_k}(y_j) \to z_j$, so that $x_j = y_j$. Therefore, $x = y$, and Σ is distal.

Next consider the direct sum $\Sigma = \Sigma_1 \oplus \Sigma_2$. A uniformity for X is obtained by taking as indices sets of the form $\alpha = \alpha_1 \cup \alpha_2$, where $\alpha_j \subseteq X_j \times X_j$ are indices. Equicontinuity of Σ follows immediately from equicontinuity of Σ_1 and Σ_2 and this observation.

Suppose that Σ_j is distal for each $j \in J$ and that $\Sigma = \bigoplus_{j \in J} \Sigma_j$. If $x \in X_j$, then $\sigma(x) = \sigma_j(x) \in X_j$, and so $\bar{O}^+(x) \subseteq X_j$. Suppose that there is a generalized sequence n_k of integers with $\sigma^{n_k}(x) \to z$ and a $y \in X$ with $\sigma^{n_k}(y) \to z$. Since the X_j are pairwise disjoint, we must have y belonging to the same X_j as x. Hence $\sigma^{n_k}(y) = \sigma_j^{n_k}(y) \to z$, $\sigma_j^{n_k}(x) \to z$, and $x, y, z \in X_j$. Since Σ_j is distal, this implies that $x = y$. Hence Σ is distal.

Now suppose that $\psi\colon \Sigma_1 \to \Sigma$ is an injection. If the mappings σ^n $(n \in Z)$ are equicontinuous on X, they are certainly equicontinuous on $\psi(X_1)$. Since ψ is a homeomorphism of X_1 onto $\psi(X_1)$, it follows that Σ_1 is equicontinuous when Σ is.

Assume that Σ is distal. If $\psi\colon \Sigma_1 \to \Sigma$ is an injection, then $\psi \otimes \psi\colon \Sigma_1 \otimes \Sigma_1 \to \Sigma \otimes \Sigma$ is also an injection. Moreover, for any $(x, y) \in X_1 \times X_1$, we have $\bar{O}(\psi(x), \psi(y)) = (\psi \times \psi)\, \bar{O}(x, y)$. Suppose $x \neq y$. Then $\psi(x) \neq \psi(y)$. Since $\bar{O}(\psi(x), \psi(y))$ is disjoint from the diagonal, so also is $\bar{O}(x, y)$. That is, Σ_1 is distal. ▮

Remarks It is also true that factors of distal systems are distal. However, the proof is surprisingly nontrivial and will be given in Section 6 using the notion of the Ellis semigroup of Σ.

Infinite direct sums of equicontinuous systems need not be equicontinuous. (See Exercise 14.)

5. INVERSE LIMITS

We turn now to the construction of inverse limits of dynamical systems. This construction generalizes that of products of infinitely many systems. We shall give a categorical definition for limits of collections of systems Σ_j indexed on a directed set J. Recall that the set J is directed by the relation $\prec$ provided that (i) $\prec$ is a partial order on J, and (ii) for each pair, $i, j \in J$ there is a $k \in J$ with $i \prec k$ and $j \prec k$.

Definition 2.11 By an *inverse system* of dynamical systems we shall mean a triple (J, Σ_j, ψ_{ij}) such that J is a directed set; for each $j \in J$, Σ_j is a dynamical system; for each pair $i, j \in J$ with $i \prec j$, the map ψ_{ij} is a homomorphism of Σ_j onto Σ_i; and the diagram

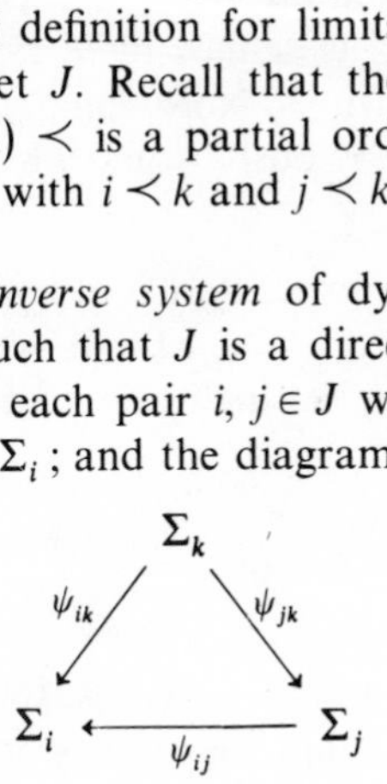

$$\begin{array}{ccc} & \Sigma_k & \\ \psi_{ik} \swarrow & & \searrow \psi_{jk} \\ \Sigma_i & \xleftarrow{\ \psi_{ij}\ } & \Sigma_j \end{array} \tag{8}$$

commutes for all $i \prec j \prec k$.

Definition 2.12 The dynamical system Σ is an *inverse limit* of the inverse system (J, Σ_j, ψ_{ij}) if (i) there exist homomorphisms ρ_j $(j \in J)$ of Σ onto Σ_j, such that all the diagrams

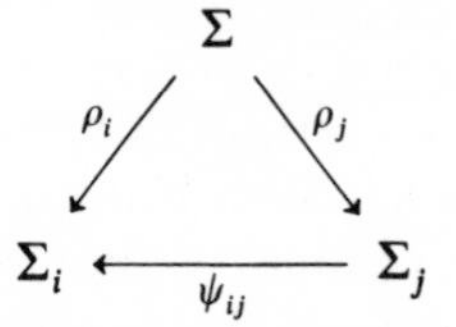

$$\begin{array}{ccc} & \Sigma & \\ {}^{\rho_i}\swarrow & & \searrow^{\rho_j} \\ \Sigma_i & \xleftarrow[\psi_{ij}]{} & \Sigma_j \end{array} \qquad (i, j \in J, i \prec j) \tag{9}$$

commute, and (ii) if Σ' is another system with epimorphisms ρ_i' satisfying (9), then there exists a homomorphism ϕ of Σ' onto Σ, such that the diagrams

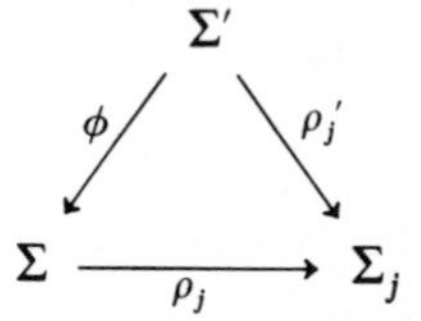

$$\begin{array}{ccc} & \Sigma' & \\ {}^{\phi}\swarrow & & \searrow^{\rho_j'} \\ \Sigma & \xrightarrow[\rho_j]{} & \Sigma_j \end{array} \qquad (j \in J) \tag{10}$$

all commute. In this case, we write $\Sigma = \text{inv lim}_{j \in J} \Sigma_j$.

Remark It follows easily from the corresponding fact in topological spaces that the inverse limit of an inverse system of dynamical systems always exists. In fact, we can take X to be the inverse limit set

$$X_\infty = \{x \in \mathop{\times}_{j \in J} X_j : \psi_{ij} x_j = x_i \text{ for all } i, j \in J, i \prec j\} \tag{11}$$

with the relative topology induced on X_∞ by the product topology on $\times_{j \in J} X_j$. Moreover, $\sigma = \sigma_\infty$ is defined by $\sigma_\infty(x) = y$, where $y_j = \sigma_j(x_j)$, and the ρ_j are the obvious projections. Note that Σ_∞ is a subsystem of $\bigotimes_{j \in J} \Sigma_j$. Any other inverse limit is obviously isomorphic to $\Sigma_\infty = (X_\infty, \sigma_\infty)$.

Examples The classic example and prototype of the inverse limit is the direct product. Indeed, if $\Sigma = \bigotimes_{i \in I} \Sigma_i$, if we let J be the collection of finite subsets of I ordered by set inclusion, and if we denote

$$\Lambda_j = \Lambda_{(i_1, \ldots, i_n)} = \Sigma_{i_1} \otimes \cdots \otimes \Sigma_{i_n} \qquad \text{for} \quad j = (i_1, \ldots, i_n) \in J,$$

then $\Sigma = \text{inv lim}_{j \in J} \Lambda_j$.

Example 7 (*Natural extension*) Suppose that $\Sigma = (X, \sigma)$ is a dynamical system, where σ is epic, but not necessarily invertible. For each positive integer n let $X_n = X$ and define $\psi_{nm} = \sigma^{m-n}$ for $m > n$. The system

$\Sigma_\infty = \text{inv lim}_{n\to\infty} \Sigma_n$ is the smallest invertible dynamical system containing Σ as a factor. To see that Σ_∞ is invertible, note that

$$X_\infty = \{x \in \mathop{\times}_{n=1}^{\infty} X_n : x_n = \sigma(x_{n+1}) \text{ for each } n\},$$

$$\sigma_\infty(x_1, x_2, x_3, \ldots) = (\sigma(x_1), x_1, x_2, \ldots),$$

so that

$$\sigma_\infty^{-1}(y_1, y_2, y_3, \ldots) = (y_2, y_3, y_4, \ldots).$$

Proposition 2.7 *Inverse limits of* (1) *minimal,* (2) *equicontinuous, or* (3) *distal dynamical systems are, respectively,* (1) *minimal,* (2) *equicontinuous, or* (3) *distal.*

Proof The statements regarding equicontinuous and distal systems follow immediately from Proposition 2.6 and the observation that Σ_∞ is a subsystem of $\bigotimes_{j\in J} \Sigma_j$.

Suppose that $\Sigma_\infty = \text{inv lim}_{j\in J} \Sigma_j$, and each Σ_j is minimal. Let A be a closed subset of X_∞ with $\sigma_\infty(A) \subseteq A$. Then for each $j \in J$, $\rho_j(A)$ is closed and $\sigma_j(\rho_j(A)) = \rho_j(\sigma(A)) \subseteq \rho_j(A) \Rightarrow \rho_j(A) = \varnothing$ or X_j. If $\rho_j(A) = X_j$ for all $j \in J$, then $A = X$. If $\rho_j(A) = \varnothing$ for any $j \in J$, then $A = \varnothing$. ∎

6. THE ELLIS SEMIGROUP OF Σ

Let $\Sigma = (X, \sigma)$ be a dynamical system. The topological space X^X of all functions from X to X (not necessarily continuous) is a compact Hausdorff space with the product topology. It is also a semigroup, with composition of functions as the operation. Clearly, $\sigma^n \in X^X$ for each $n = 0, 1, 2, \ldots$.

Definition 2.13 The *Ellis semigroup* $E(\Sigma)$ of the dynamical system $\Sigma = (X, \sigma)$ is the closure in X^X of the semigroup $T(\Sigma) = \{\sigma^n : n = 0, 1, 2, \ldots\}$.

We note first that $E(\Sigma)$ is compact Hausdorff and is a semigroup. The latter follows from the following observations. Let $E_0(\Sigma)$ be the set of all continuous functions in $E(\Sigma)$. Thus $T(\Sigma) \subseteq E_0(\Sigma) \subseteq E(\Sigma)$. It is immediately verified that

(i) $g \to gh$ is a continuous map of $E(\Sigma)$ for each $h \in E(\Sigma)$, and
(ii) $g \to hg$ is a continuous map of $E(\Sigma)$ for each $h \in E_0(\Sigma)$.

(Recall that the topology in X^X is the topology of pointwise convergence of nets.) Now if $g, h \in E(\Sigma)$, then there exist nets $g_\alpha, h_\beta \in T(\Sigma)$ with $g_\alpha \to g$ and $h_\beta \to h$. From (ii) we have $g_\alpha h_\beta \to g_\alpha h$ (convergence in β) for fixed α, and from (i) $g_\alpha h \to gh$. Thus $gh \in E(\Sigma)$, which is therefore a semigroup. Clearly, $E_0(\Sigma)$ is also a semigroup.

Theorem 2.4 *$E(\Sigma)$ is a group iff Σ is distal.*

Proof Suppose first that $E(\Sigma)$ is a group, and that there exist $x, y, z \in X$ and a net n_k of integers such that $\sigma^{n_k}x \to z$, $\sigma^{n_k}y \to z$. By compactness of $E(\Sigma)$ we may assume, by passing to a subnet if necessary, that $\sigma^{n_k} \to g \in E(\Sigma)$. But then $g(x) = g(y) = z$, and since g is invertible, it follows that $x = y$.

Conversely, suppose that Σ is distal. Then each $g \in E(\Sigma)$ is monic. For if $\sigma^{n_k} \to g$ and $g(x) = g(y)$, then by distality $x = y$. It follows immediately that $E(\Sigma)$ has a left cancellation law:

$$gg_1 = gg_2 \Rightarrow g_1 = g_2.$$

In particular, the only idempotent $g^2 = g$ in $E(\Sigma)$ is the identity e.

Let $h \in E(\Sigma)$. We shall show that h has a left inverse. Let $E_1 = \{gh : g \in E(\Sigma)\}$. Then $E_1{}^2 \subseteq E_1$, that is E_1 is a subsemigroup of $E(\Sigma)$. According to (i) above, E_1 is closed [since $E(\Sigma)$ is compact].

Let $\mathcal{S}$ be the collection of all closed, nonempty subsets S of E_1 such that $S^2 \subseteq S$. Since $E_1 \in \mathcal{S}$, it is a nonempty collection. Let $\mathcal{S}_1$ be a decreasing chain in $\mathcal{S}$. By compactness of E_1, $S_1 = \cap\, \mathcal{S}_1$ is nonempty. Moreover, S_1 is closed, and $S_1{}^2 \subseteq S_1$. By Zorn's lemma $\mathcal{S}$ must contain a minimal element S_0.

Let $g \in S_0$. We shall show that $g^2 = g$. Since $S_0 g = \{fg : f \in S_0\} \in \mathcal{S}$ and S_0 is minimal, we have $S_0 g = S_0$. Thus $g = fg$ for some $f \in S_0$. Now $W = \{p \in S_0 : g = pg\}$ is nonempty, since $f \in W$. According to (i), W is closed since it is the inverse image of $\{g\}$ under the continuous mapping $p \to pg$. Since also $W^2 \subseteq W$, we have $W \in \mathcal{S}$, and again by the minimality of S_0, $W = S_0$. This means that $g \in W$, and $g = g^2$ as asserted. From above we must have $g = e \in E_1$; that is, h has a left inverse. Since $h \in E(\Sigma)$ was arbitrary, $E(\Sigma)$ is a group. ∎

Remark We have assumed up to now that distal systems were invertible. The above proof that $E(\Sigma)$ is a group does not make use of this assumption but yields invertibility of σ as a byproduct. (See Exercise 18.)

Corollary 2.4.1 *Every distal system is semisimple.*

Proof Suppose $\Sigma = (X, \sigma)$ is distal, and let $x \in X$. Let $y \in \bar{O}(x)$, say $\sigma^{n_k}x \to y$. By passing to a subnet, we may assume that $\sigma^{n_k} \to g \in E(\Sigma)$. Thus $g(x) = y$. Since $E(\Sigma)$ is a group, we have $x = g^{-1}(y)$ with $g^{-1} \in E(\Sigma)$. Thus $x \in \bar{O}(y)$. It follows that $\bar{O}(x) = \bar{O}(y)$ is a minimal set. Thus X is a disjoint union of minimal sets for σ, and Σ is semisimple. ▮

Now suppose that $\Sigma_1 = (X_1, \sigma_1)$ is a factor of the dynamical system $\Sigma = (X, \sigma)$, say $\psi\colon \Sigma \to \Sigma_1$. Then ψ induces a semigroup homomorphism ψ^* of $E(\Sigma)$ onto $E(\Sigma_1)$ as follows.

Suppose $g \in E(\Sigma)$, and let g_α, h_α be nets in $T(\Sigma)$ with $g_\alpha \to g$ and $h_\alpha \to g$. Let $y \in X_1$. Then $y = \psi(x)$ for some $x \in X$. Corresponding to each $g_\alpha = \sigma^{n_\alpha} \in T(\Sigma)$ there is a $g_\alpha{}^* = \sigma_1^{n_\alpha} \in T(\Sigma_1)$. Moreover,

$$g_\alpha{}^*(y) = \sigma_1^{n_\alpha} y = \sigma_1^{n_\alpha}\psi(x) = \psi(\sigma^{n_\alpha}x) = \psi(g_\alpha(x)) \to \psi(g(x)).$$

Similarly, $h_\alpha{}^*(y) = \psi(h_\alpha(x)) \to \psi(g(x))$. That is, both nets $g_\alpha{}^*$ and $h_\alpha{}^*$ converge to the same element $g^* \in E(\Sigma_1)$, and $g^*(\psi(x)) = \psi(g(x))$. Let us write $\psi^*g = g^*$. Thus ψ^*g is the unique element of $E(\Sigma_1)$ that completes the diagram

$$\begin{array}{ccc} X & \xrightarrow{\quad g \quad} & X \\ \Big\downarrow \psi & & \Big\downarrow \psi \\ X_1 & \xrightarrow{\quad \psi^* g \quad} & X_1 \end{array} \qquad (12)$$

The following observations are easily verified by examining diagram (12) and recalling that $\psi^*\sigma = \sigma_1$.

Proposition 2.8 *If $\psi^*\colon E(\Sigma) \to E(\Sigma_1)$ is defined by* (12), *then*

(i) *ψ^* is onto,*
(ii) *ψ^* is a semigroup homomorphism,*
(iii) *ψ^*g is continuous when g is continuous, and*
(iv) *ψ^* is continuous.*

Combined with Theorem 2.4 this yields the following result promised in Section 4.

Corollary 2.4.2 *Factors of distal systems are distal.*

Proof The homomorphic image of a group is a group. ▮

Proposition 2.9 *If Σ is equicontinuous, then $E(\Sigma)$ is a group of homeomorphisms.*

Proof If Σ is equicontinuous, it is distal. Hence $E(\Sigma)$ is a group. Moreover, equicontinuity of the mappings σ^{n_α} is easily seen to imply continuity of the limit function $g \in E(\Sigma)$, where $\sigma^{n_\alpha} \to g$. Since g is invertible and X is compact, g is a homeomorphism. ∎

Remarks *1* The time has come to reveal our fraud. Much more than Proposition 2.9 can be proved. First of all, (see [18]) the converse of that proposition is true. Moreover, using a result from [17] and observations (i) and (ii) following Definition 2.13, it can be shown that $E(\Sigma)$ is a topological group when Σ is equicontinuous; that is, multiplication is jointly continuous. If, in addition, Σ is minimal, then $E(\Sigma)$ is homeomorphic to X [map g to $g(x_0)$ for some fixed $x_0 \in X$] and the inverse homeomorphism ψ^{-1} carries the action of σ on X to multiplication by σ in $E(\Sigma)$. That is,

$$\psi(\sigma g) = g(\sigma x_0) = \sigma(\psi(g)). \tag{13}$$

Thus *every equicontinuous dynamical system is a disjoint union (sum) of systems each of which is given by a rotation of a compact group.* Such rotations are special cases of the affine transformations to be discussed in Chapter III.

2 The situation for distal systems is much more complex. However, Furstenberg [22] has given a complete structure theorem in this case as well. We shall content ourselves here by stating (without proof) a related characterization of minimal distal systems from [23]. First we must define group extensions.

Definition 2.14 Let $\Sigma = (X, \sigma)$ be a dynamical system, and let G be a group of homeomorphisms of X. We assume that

(i) $(g, x) \to g(x)$ is jointly continuous for $g \in G$, $x \in X$,
(ii) $(gh)(x) = g(h(x))$ for all $g, h \in G$, $x \in X$,
(iii) $g(x) = x$ for some $x \in X \Rightarrow g$ is the identity of G, and
(iv) $g(\sigma x) = \sigma g(x)$ for all $g \in G$, $x \in X$.

The orbit space X/G, whose typical element is a set of the form $Gx = \{g(x) : g \in G\}$ for some $x \in X$, is compact in the quotient topology and supports a dynamical system $\Sigma/G = (X/G, \hat{\sigma})$, where $\hat{\sigma}(Gx) = G(\sigma x)$. Σ/G is a factor of Σ, and Σ is said to be a *group extension* of Σ/G.

We remark in passing that, if Σ is equicontinuous, then $G = E(\Sigma)$ is such a group. In this case, the orbit space consists of isolated points, the minimal sets for σ, and $\hat{\sigma}$ is the identity. Thus every minimal equicontinuous system is a group extension of the trivial (one-point) system.

Theorem (*Furstenberg*) *The class of minimal distal systems is the smallest class $\mathscr{C}$ of dynamical systems satisfying:*

(a) *the trivial system is in $\mathscr{C}$;*
(b) *factors of systems in $\mathscr{C}$ are in $\mathscr{C}$;*
(c) *inverse limits of systems in $\mathscr{C}$ are in $\mathscr{C}$;*
(d) *group extensions of systems in $\mathscr{C}$ are in $\mathscr{C}$ provided they are minimal.*

3 For a detailed account of the algebraic approach to dynamical systems begun in this section the reader is referred to the monograph [18] by Ellis.

7. EXPANSIVE SYSTEMS

The motivation for the concept introduced in this section is the symbolic or shift dynamical system of Example 1. In many regards, expansive systems lie at the opposite end of the spectrum from distal systems.

We shall assume in this section that Σ is invertible.

Definition 2.15 The (invertible) dynamical system $\Sigma = (X, \sigma)$ is said to be *expansive* if there exists an index α on X (the expansive index) such that for all $x, y \in X$ with $x \neq y$ there is an integer n with $(\sigma^n x, \sigma^n y) \notin \alpha$.

Example If the X of Example 1 is endowed with the metric

$$d(x, y) = (1 + \min\{|n| : x_n \neq y_n\})^{-1},$$

then an expansive index for Σ is given by

$$\alpha = \{(x, y) : d(x, y) < 1\}.$$

Remarks *1* If α is an expansive index, then clearly any $\beta \subseteq \alpha$ is also. However, the above example exhibits a maximal such index.

2 If Σ is expansive, it cannot be equicontinuous. In fact (Exercise 23), it cannot even be distal.

3 It is clear that subsystems and finite sums of expansive systems are expansive. It is not obvious that factors are. On the other hand, the principal structure theorem below says that all expansive systems can be obtained as factors of subsystems of the symbolic systems of Example 1. This result was obtained independently by Keynes and Robertson [40] and by Reddy [50].

We need to define the notion of a "generator" for the dynamical system Σ (so called because of its role in the calculation of topological entropy).

Definition 2.16 Let $\Sigma = (X, \sigma)$ be an invertible dynamical system. Let $\mathscr{U}$ be a finite open cover of X. Then $\mathscr{U}$ is a *generator* for Σ if $\bigcap_{n=-\infty}^{\infty} \sigma^{-n}(\overline{A}_n)$ is either empty or a single point for each choice of $A_n \in \mathscr{U}$ $(n \in Z)$.

Example If Σ is the symbolic system on k points of Example 1, we can take $\mathscr{U} = \{U_0, U_1, \ldots, U_{k-1}\}$, where $U_j = \{x \in X : x_0 = j\}$. Then $\overline{A}_n = A_n$ for each $A_n \in \mathscr{U}$ and $\bigcap_{n=-\infty}^{\infty} \sigma^{-n}(\overline{A}_n)$ contains exactly one point, namely $x = \{x_n\}$, where $x_n = j$ if $A_n = U_j$.

Clearly, $\mathscr{U} \cap Y = \{U \cap Y : U \in \mathscr{U}\}$ is a generator for the subsystem $\Sigma_1 = (Y, \sigma)$ of Σ when $\mathscr{U}$ is a generator for Σ.

Proposition 2.10 *Σ has a generator iff it is expansive.*

Proof Suppose Σ is expansive with expansive index α. Choose a symmetric open index β on X with $\beta^2 \subseteq \alpha$, that is, a β such that

$$(x, y) \in \beta \Rightarrow (y, x) \in \beta$$

$$(x, y) \in \beta, (y, z) \in \beta \Rightarrow (x, z) \in \alpha.$$

Since X is compact, there is a finite cover $\mathscr{U}$ of X consisting of sets of the form $x\beta = \{y : (x, y) \in \beta\}$ (β-neighborhoods). Suppose that there exists a choice of sets $A_n \in \mathscr{U}$ $(n \in Z)$ such that $\bigcap_{n=-\infty}^{\infty} \sigma^{-n}(A_n)$ contains two distinct points x and y. If $A_n = x_n\beta$, then

$$(\sigma^n x, \sigma^n y) \in (x_n\beta) \times (x_n\beta) \subseteq \beta^2 \subseteq \alpha$$

for each $n \in Z$. But this contradicts the assumption that α is an expansive index. Thus $\bigcap_{n=-\infty}^{\infty} \sigma^{-n}(A_n)$ contains at most one point for each choice of $A_n \in \mathscr{U}$ $(n \in Z)$.

In general, $\mathscr{U}$ will not be a generator, but we shall make use of it to construct one. For each $x \in X$, let $x \in A_x \in \mathscr{U}$. Choose an open neighborhood V_x of x with $\overline{V}_x \subseteq A_x$. The sets V_x cover X and we can choose a finite subcover $\mathscr{W}$. For each choice of $B_n \in \mathscr{W}$ $(n \in Z)$, $\overline{B}_n \subseteq A_n \in \mathscr{U}$ and so $\bigcap_{n=-\infty}^{\infty} \sigma^{-n}(\overline{B}_n)$ contains at most one point. Thus $\mathscr{W}$ is a generator.

Conversely, suppose that $\mathscr{U}$ is a generator for Σ. Then there exists an index α on X (called the Lebesgue index of $\mathscr{U}$) such that for each $x \in X$ there is some $U \in \mathscr{U}$ that contains $x\alpha$. Suppose that $(\sigma^n x, \sigma^n y) \in \alpha$ for each $n \in Z$. Choose $A_n \in \mathscr{U}$ such that $(\sigma^n x)\alpha \subseteq A_n$. Then $\sigma^n y \in A_n$ and, clearly, $\sigma^n x \in A_n$. It follows that both x and y belong to $\bigcap_{n=-\infty}^{\infty} \sigma^{-n}(A_n)$. Since $\mathscr{U}$ is a generator, $x = y$. ∎

The existence of a generator for Σ turns out to be exactly what is needed for an efficient "coding" of the sequence $\{\sigma^n x\}$ for $x \in X$. Indeed, if $\mathscr{U}$ is a generator for Σ, say $\mathscr{U} = \{U_0, U_1, \ldots, U_{k-1}\}$, then we can define a mapping ψ of a subset Z of $Y = \times_{n=-\infty}^{\infty} \{0, 1, \ldots, k-1\}$ into X as follows. Let Z be the set of sequences $y = \{y_n\} \in Y$ such that $\bigcap_{n=-\infty}^{\infty} \sigma^{-n}(\overline{U}_{y_n})$ is nonempty, and define $\psi(y)$ for $y \in Z$ to be the unique element x belonging to that intersection.

Proposition 2.11 *Z is a closed shift-invariant subset of Y, and the mapping $\psi: Z \to X$ is continuous and epic.*

Proof Since $\mathscr{U}$ is a cover, for each $x \in X$ there exists (possibly more than) one sequence $\{y_n\}$ such that $\sigma^n x \in U_{y_n}$ $(n \in Z)$. Thus ψ is epic. Moreover, if $\bigcap_{n=-\infty}^{\infty} \sigma^{-n}(\overline{U}_{y_n}) \neq \varnothing$, then

$$\bigcap_{n=-\infty}^{\infty} \sigma^{-n}(\overline{U}_{y_{n+1}}) = \bigcap_{n=-\infty}^{\infty} \sigma^{-n+1}(\overline{U}_{y_n}) = \sigma\left(\bigcap_{n=-\infty}^{\infty} \sigma^{-n}(\overline{U}_{y_n})\right) \neq \varnothing,$$

so that Z is shift invariant. That is, $\tau(Z) = Z$, where τ is the shift operator on Y defined by $\tau(y) = w$, with $w_n = y_{n+1}$.

Now let $y^j = \{y_n^j\}$ be a sequence in Z. Thus there exists a sequence $x_j \in X$ such that

$$\bigcap_{n=-\infty}^{\infty} \sigma^{-n}(\overline{U}_{y_n^j}) = \{x_j\}. \tag{14}$$

Suppose that $y^j \to y \in Y$. By passing to a subnet if necessary, we may assume that $x_j \to x \in X$. For each $n \in Z$ the net of integers y_n^j converges to the integer y_n. It follows that $y_n^j = y_n$ for $j \geq j(n)$. Thus $\sigma^n x_j \in \overline{U}_{y_n}$ for $j \geq j(n)$, and, by continuity of σ^n, $\sigma^n x \in \overline{U}_{y_n}$. That is,

$$x \in \bigcap_{n=-\infty}^{\infty} \sigma^{-n}(\overline{U}_{y_n}) \neq \varnothing,$$

so that Z is closed and

$$\bigcap_{n=-\infty}^{\infty} \sigma^{-n}(\overline{U}_{y_n}) = \{x\}. \tag{15}$$

Since (15) holds for any cluster point of the sequence $\{x_j\}$, it follows from (14) and (15) that $x_j = \psi(y^j) \to x = \psi(y)$. Thus ψ is continuous. ∎

Theorem 2.5 (*Keynes–Robertson–Reddy*) *Let Σ be an expansive dynamical system. Then Σ is a factor of a subsystem of a symbolic system.*

Proof It only remains to show that $\psi\colon (Z, \tau) \to \Sigma$ is a homomorphism, that is, that $\psi\tau = \sigma\psi$. But this is a routine verification:

$$\{\sigma\psi(y)\} = \sigma\left[\bigcap_{n=-\infty}^{\infty} \sigma^{-n}(\bar{U}_{y_n})\right] = \bigcap_{n=-\infty}^{\infty} \sigma^{-n}(\bar{U}_{y_{n+1}}) = \{\psi(\tau y)\}. \quad \blacksquare$$

Corollary 2.5.1 *If $\Sigma = (X, \sigma)$ is expansive, then X is metrizable.*

The homomorphism ψ of Theorem 2.5 is not in general an isomorphism, so the coding referred to above is not well defined. However, it can be shown to be so in case X is zero-dimensional [40, 50]. In this case, of course, the converse of the theorem also holds (with "factor of a" deleted).

We conclude this section by noting that W. Krieger has shown [42] that every ergodic invertible *abstract* dynamical system with finite entropy (see Chapter IV) is *measure-theoretically* isomorphic to a subsystem of a symbolic system, hence to an expansive system. Moreover, the subsystem may be chosen to be uniquely ergodic.

EXERCISES

Minimality and Ergodicity

1. (a) Discuss the "adding machine" transformation ϕ (Exercise 5 of Chapter I) as a transformation of a compact (totally disconnected) Hausdorff space. (Use the product topology.)

(b) Let μ be *any* invariant Borel measure for this ϕ. Show that

$$\phi^{-a}\{x : x_1 = a\} = \{x : x_1 = 0\} \qquad (a = 0, 1, \ldots, k_1),$$

and so

$$\mu\{x : x_1 = a\} = \frac{1}{k_1 + 1} \qquad (a = 0, 1, \ldots, k_1).$$

(c) Show that there exists an integer p depending on $a_1, \ldots, a_n$ such that

$$\phi^{-p}\{x : x_1 = a_1, \ldots, x_n = a_n\} = \{x : x_1 = 0, \ldots, x_n = 0\}$$

and conclude that this set must have measure

$$\frac{1}{k_1 + 1} \cdot \frac{1}{k_2 + 1} \cdots \frac{1}{k_n + 1}.$$

This proves that μ is the measure constructed in Chapter I, and that ϕ is uniquely ergodic.

2. Show that the symbolic dynamical system of Example 1 is not uniquely ergodic.

3. If (X, σ) is invertible and uniquely ergodic, then $(X, \mathscr{B}, \mu, \sigma)$ is ergodic, where μ is the unique σ-invariant measure.

4. Let $\Sigma = (X, \sigma)$ be an invertible dynamical system. A point $x \in X$ is *almost periodic* if for each open neighborhood U of x the set $A_x(U) = \{n \in Z : \sigma^n(x) \in U\}$ has "bounded gaps"; that is, there exists an integer k such that every set of k successive integers contains at least one element of $A_x(U)$.

(a) If $x \in X_0 \subseteq X$ is such that $\Sigma_0 = (X_0, \sigma)$ is minimal, and if U is a neighborhood of x, show as in the proof of Theorem 2.2 that $O(x) \subseteq \bigcup_{j=1}^{k} \sigma^{n_j}(U)$. Deduce that the gaps in $A_x(U)$ are bounded by $k = \max_j n_j - \min_j n_j$. Thus every invertible dynamical system has an almost periodic point, and each point is almost periodic for a minimal invertible system.

(b) Show conversely that if x is almost periodic, then $\bar{O}(x)$ is a minimal invariant set. [Show $y \in O(x) \subseteq \bigcup_{j=1}^{k} \sigma^{n_j}(U)$ implies $O(y) \cap U \neq \emptyset$, and hence that $x \in O(y)$.]

5. **(a)** Referring to Example 1, show that $x \in X$ is an almost periodic point iff for each integer $n \geq 1$ there exists a set A_n with bounded gaps such that

$$x_{i+j} = x_i \qquad (-n \leq i \leq n, \ \ j \in A_n). \tag{16}$$

(b) For the x of Example 4 and for $n = 0$, show that (16) holds, with the gaps in $A_0 = \{j : x_j = x_0 = 0\}$ bounded by 3.

(c) It can be shown [28, p. 107] that, in general, (16) holds with the gaps in A_n bounded by 2^{n+5}. Deduce that $\Sigma_0 = (\bar{O}(x), \sigma)$ is minimal.

(d) Find the smallest positive $j \in A_n$ for $n = 0, 1, 2, 3, 4$.

6. If $x \in X$ is a generic point for $\Sigma = (X, \sigma)$, that is, if the limit

$$L_x(f) = \lim_{n\to\infty} \frac{1}{n} \sum_{k=0}^{n-1} f(\sigma^k(x))$$

exists for each $f \in C(X)$, show that L_x is a positive linear functional on $C(X)$ with $L_x(1) = \|L_x\| = 1$ and $T_\sigma{}^* L_x = L_x$. Hence there exists a unique normalized Borel measure μ_x on X such that

$$L_x(f) = \int_X f \, d\mu_x \qquad (f \in C(X)).$$

7. **(a)** Suppose that $\Sigma = (X, \sigma)$ is a minimal dynamical system, and that μ is a σ-invariant Borel measure on X. If U is an open set, show that there exist positive integers $n_1, n_2, \ldots, n_l$ such that $X \subseteq \bigcup_{j=1}^{l} \sigma^{-n_j}(U)$, and hence that $\mu(U) > 0$. This means that the support of μ is X.

(b) If Σ is not minimal, there exists a point $x \in X$ with $\bar{O}^+(x) \neq X$. Show that the measure μ_x defined as in Exercise 6 by

$$\int_X f \, d\mu_x = \lim_{n \to \infty} \frac{1}{n} \sum_{k=0}^{n-1} f(\sigma^k(x)) \qquad (f \in C(X))$$

has its support contained in $\bar{O}^+(x)$. (Find a continuous $f \geq 0$ which vanishes on $\bar{O}^+(x)$ and equals 1 on the compact set $A \subseteq X \sim \bar{O}^+(x)$. Use regularity of μ_x to conclude that $\mu_x(X \sim \bar{O}^+(x)) = 0$.)

8. **(a)** Show that the set $M(\Sigma) \subseteq M(X)$ of normalized σ-invariant Borel measures on X is weak*-compact and convex.

(b) A point α of a convex set M is an *extreme point* if it cannot be expressed as a nontrivial convex combination of two distinct points of M. Show that the extreme points of $M(\Sigma)$ are the ergodic invariant measures.

(c) Show (by the Krein–Milman theorem [16, p. 440]) that a dynamical system Σ always has an ergodic invariant measure, and that Σ is uniquely ergodic iff there is only one ergodic invariant measure.

9. Consider the following assertions about the invertible dynamical system $\Sigma = (X, \sigma)$:

(1) Σ is minimal;

(2) Σ is ergodic, that is, the set of points $x \in X$ for which $\bar{O}(x) \neq X$ is of first category;

(3) Σ is uniquely ergodic;

(4) $\bar{O}(x) = X$ for some $x \in X$;

(5) there exists an ergodic σ-invariant Borel measure μ with support X;

(6) $f \in C(X)$, $T_\sigma f = f \Rightarrow f = \text{constant}$;

(7) $f \in B(X)$, $T_\sigma f = f \Rightarrow f = \text{constant}$, where $B(X)$ is the space of all bounded functions on X.

(a) Show that the following implications hold:

$$\begin{array}{c} (1) \Rightarrow (5) \Rightarrow (2) \Rightarrow (4) \Rightarrow (6) \\ \qquad\qquad\qquad\qquad \nearrow \\ (7) \Rightarrow (3) \end{array}$$

(b) Show that the shift dynamical system (Example 1) on two points satisfies (4) but not (3).

(c) Show that the transformation σ defined on the unit circle $K = \{z : |z| = 1\}$ by $\sigma(e^{2\pi i t}) = \exp(2\pi i t^2)$ satisfies (3) but not (4).

Equicontinuity and Distality

10. Show that the following assertions are pairwise equivalent.

(1) $\Sigma = (X, \sigma)$ is distal.

(2) For each $x \neq y$ there exists an index α on X with $(\sigma^n(x), \sigma^n(y)) \notin \alpha$ for all n.

(3) $\overline{O}_{\sigma^2}((x, y)) \cap \Delta = \varnothing$ for $x \neq y$.

(4) $\{(x, y) : \forall$ index α $\exists$ integer n with $(\sigma^n(x), \sigma^n(y)) \in \alpha\} = \Delta$.

(5) $\sigma^{n_k}(x) \to z$, $\sigma^{n_k}(y) \to z$ for some generalized sequence n_k of integers $\Rightarrow x = y$.

11. Show that the system in Example 5 is minimal.

12. Verify the details of Example 6.

Sums, Products, Subsystems, and Factors

13. **(a)** Prove Proposition 2.5.

(b) Show that nontrivial direct sums are *never* minimal. Show that $\Sigma \otimes \Sigma$ is not minimal.

14. Give an example of an infinite direct sum $\Sigma = \bigoplus_{j \in J} \Sigma_j$ such that each Σ_j is equicontinuous, but Σ is not.

15. **(a)** Suppose that $\psi \colon \Sigma_2 \to \Sigma_1$. Let $M(\Sigma_i)$ and $M(X_i)$ be defined as in Exercise 8. For each $\nu \in M(\Sigma_1)$ show that

$$A = \{\mu \in M(X_2) : T_\psi{}^*\mu = \nu, \mu \geq 0, \mu(X_2) = 1\}$$

is a weak*-closed, convex, nonempty set, and that $T^*_{\sigma_2}(A) \subseteq A$. Hence deduce that there exists a $\mu \in M(\Sigma_2)$ with $T_\psi{}^*\mu = \nu$.

(b) Show that a factor of a uniquely ergodic dynamical system is uniquely ergodic. How about products, sums, subsystems of uniquely ergodic systems?

Inverse Limits

16. Show that the system Σ_∞ described in the Remark following Definition 2.13 is an inverse limit of (J, Σ_j, ψ_{ij}), and that any other inverse limit is isomorphic to Σ_∞.

17. Show that an infinite direct product is the inverse limit of finite direct products.

The Ellis Semigroup

18. Give a "one-sided" definition of distality for noninvertible systems. Show that any such system must in fact be invertible and distal in the former sense.

19. Prove Proposition 2.8.

20. Show that Σ is distal iff $\Sigma \otimes \Sigma$ is semisimple. Use this to obtain another proof of Corollary 2.4.2.

21. If Σ is minimal and equicontinuous, show that $g(x_0) = x_0$ for some $x_0 \in X$ implies $g(\sigma^n x_0) = \sigma^n x_0$ for each n and hence that $g(x) = x$ for all $x \in X$. Conclude that the mapping ψ defined by $\psi(g) = g(x_0)$ is an isomorphism of $\hat{\Sigma} = (E(\Sigma), \hat{\sigma})$ onto $\Sigma = (X, \sigma)$, where $\hat{\sigma}(g) = \sigma g$.

22. **(a)** Let σ be rotation through an irrational multiple of π on the circle $K = \{z : |z| = 1\}$. Show by direct calculation that $E(K, \sigma)$ is homeomorphic to K.

(b) Let σ be defined on K by

$$\sigma(e^{2\pi ix}) = \exp(2\pi ix^2).$$

Show that $E(K, \sigma) = T(K, \sigma) \cup \{g\}$, where g is the constant function $g(z) = 1$.

(c) Let Σ be as in Example 5. Show that $E(\Sigma)$ consists of all functions g of the form

$$g(z, w) = (\lambda z, h(z)w),$$

where λ is an arbitrarily complex number of absolute value one, and h is any (not necessarily continuous) function from K to K satisfying

$$h(e^{2\pi ia}z) = \lambda h(z).$$

Expansive Systems

23. Suppose $\Sigma = (X, \sigma)$ is expansive. Show that there exist points $x, y \in X$ that are positively asymptotic, that is, such that $\lim_{n \to +\infty} \rho(\sigma^n x, \sigma^n y) = 0$, where ρ is a metric yielding the topology of X. Conclude that Σ is not distal.

24. Show by example that neither ψ nor ψ^{-1} necessarily carries generators onto generators, where $\psi\colon \Sigma \to \Sigma_1$ is a homomorphism.

25. Prove Corollary 2.5.1 by showing that a continuous map from a second-countable compact space onto a Hausdorff space has a second-countable image.

CHAPTER

III

Group Automorphisms and Affine Transformations

1. DYNAMICAL SYSTEMS ON GROUPS

In this chapter we particularize the general results of the previous chapters to the study of a continuous affine transformation ϕ of a compact abelian group G. The system (G, ϕ) is a classical dynamical system in the sense of Chapter II and with the Haar measure structure on G becomes an abstract dynamical system such as was studied in Chapter I. Thus the results of the first two chapters are combined in a natural way.

Moreover, as we shall see, much more complete results are possible in terms of structure and representation theorems than is generally true. This is partially due to the fact that the duality theory for locally compact groups permits us to study analytic properties of such systems by algebraic methods on discrete (nontopological) groups, such as in the beautiful result of Seethoff [56] concerning automorphisms with zero entropy (Theorem 4.11 of Chapter IV). On the other hand, the now-classical results of Abramov [2] and of Hahn and Parry [29] [Theorem 3.9(ii)] concerning systems with quasidiscrete spectrum indicate the degree of generality attained in studying dynamical systems on groups. (See also Theorem 2.5 and Remark 1 following Proposition 2.9.)

Throughout this chapter, we shall let G denote a compact abelian (Hausdorff) group. We shall denote the group operation by $+$ and the identity element of G by 0, except in certain examples where the multiplicative notation is more natural. The *dual group* (or character group) of G will be denoted by $\hat{G}$ and the value of the character $\gamma \in \hat{G}$ at $x \in G$ by $\langle x, \gamma \rangle$. Since G is compact, $\hat{G}$ is discrete, that is, an abstract group. The normalized (total measure one) Haar measure on G will be denoted by m and the Fourier transform of a function $f \in L_1(G, m)$ by $\hat{f}$. Thus

$$\hat{f}(\gamma) = \int_G f(x)\overline{\langle x, \gamma \rangle} m(dx). \tag{1}$$

For further explanation of these terms as well as the basic theorems we will need about topological groups, the reader is referred, for example, to the first two chapters of the monograph [53].

Let τ: $G \to G$ be a continuous automorphism (or epimorphism). It is easily demonstrated that τ is a Haar measure-preserving transformation (Exercise 1). Likewise, for each $a \in G$ the translation $x \to x + a$ $(x \in G)$ is a measure-preserving transformation. It follows that the formula

$$\phi(x) = \tau(x) + a \tag{2}$$

defines a continuous function ϕ: $G \to G$ which, as the composition of two measure-preserving transformations, is measure preserving. A function of the form (2) is said to be an *affine transformation* of G.

Example 1 The simplest nonfinite compact group is the unit interval $G = \{x : 0 \leq x < 1\}$ with addition modulo one or, equivalently, the *circle group* $K = \{z \in C : |z| = 1\}$ in the complex plane, with complex multiplication as the group operation. The correspondence between G and K is given by $x \leftrightarrow e^{2\pi i x}$. The dual $\hat{G}$ of G is the additive group Z of integers, with $\langle x, n \rangle = e^{2\pi i n x}$.

The only continuous automorphisms of G are the identity $x \to x$ and the map $x \to 1 - x$. Thus the affine transformations of G are of the form $x \to x + a$ or $x \to a - x$. On the other hand, for each integer $n \in Z$ the map $x \to nx$ modulo one is an epimorphism of G.

We define the *adjoint* τ^*: $\hat{G} \to \hat{G}$ of a continuous endomorphism τ: $G \to G$ by $\langle x, \tau^*(\gamma) \rangle = \langle \tau(x), \gamma \rangle$ $(x \in G, \gamma \in \hat{G})$. Then τ^* is an endomorphism, which is epic (monic) iff τ is monic (epic).

Example 2 The *n-torus* K^n is the product of n copies of the circle group K or G of the previous example. Let us adopt the additive notation as in G of that example. Thus the dual of K^n is the lattice Z^n

of points in n-dimensional Euclidean space having integer coordinates. The duality is given by

$$\langle x, \gamma \rangle = \exp[2\pi i(\gamma_1 x_1 + \gamma_2 x_2 + \cdots + \gamma_n x_n)]. \tag{3}$$

Let $T = (t_{ij})$ be any $n \times n$ matrix with integer entries t_{ij} $(i, j = 1, \ldots, n)$ and nonzero determinant. Define τ on K^n by considering $x = (x_1, \ldots, x_n) \in K^n$ as a column vector and setting $\tau(x) = Tx$ modulo one. It follows that $\tau(x) \in K^n$ and that τ is an endomorphism. Indeed, τ is an epimorphism (Exercise 2). If the determinant of the integral matrix T is ± 1, it is called *unimodular*.

Proposition 3.1 *Each unimodular matrix $T = (t_{ij})$ determines an automorphism τ of K^n by $\tau(x) = Tx$ modulo one. Conversely, every continuous automorphism of K^n has such a representation.*

Proof Suppose that T is unimodular. In particular, T is nonsingular and the inverse matrix T^{-1} is also unimodular. Thus $\sigma(x) = T^{-1}x$ modulo one defines an endomorphism of K^n. Thus for $x \in K^n$ we have $T^{-1}x = \sigma(x) + y$, where $y = (y_1, \ldots, y_n)$ has only integer components. It follows that $x = T(T^{-1}x) = T(\sigma(x)) + Ty = \tau(\sigma(x))$ modulo one. Similarly, $\sigma(\tau(x)) = x$ for all $x \in K^n$. Hence $\sigma = \tau^{-1}$, and τ is an automorphism.

Conversely, suppose that τ is an automorphism of K^n. Let τ^* denote the adjoint of τ on Z^n. For each $j = 1, 2, \ldots, n$ let $\delta^i \in Z^n$ be the vector whose jth component is δ_{ij}. Define t_{ij} to be the jth component of $\tau^*(\delta^i)$. For each $\gamma \in Z^n$ we have

$$\begin{aligned}
\tau^*(\gamma) &= \tau^*(\gamma_1 \delta^1 + \cdots + \gamma_n \delta^n) = \gamma_1 \tau^*(\delta^1) + \cdots + \gamma_n \tau^*(\delta^n) \\
&= \gamma_1(t_{11}, \ldots, t_{1n}) + \cdots + \gamma_n(t_{n1}, \ldots, t_{nn}) \\
&= (\gamma_1 t_{11} + \cdots + \gamma_n t_{n1}, \ldots, \gamma_1 t_{1n} + \cdots + \gamma_n t_{nn}) \\
&= \gamma T.
\end{aligned} \tag{4}$$

Now define $\sigma\colon K^n \to K^n$ by $\sigma(x) = Tx$ modulo one. For $x, \gamma \in R^n$ let us set $\gamma \cdot x = \gamma_1 x_1 + \cdots + \gamma_n x_n$. For fixed $x \in K^n$ let $y = \sigma(x)$. Then for each $\gamma \in Z^n$ we have

$$\begin{aligned}
\langle \sigma(x), \gamma \rangle &= e^{2\pi i \gamma \cdot y} = e^{2\pi i \gamma \cdot (Tx)} = e^{2\pi i (\gamma T) \cdot x} \\
&= \langle x, \tau^*(\gamma) \rangle = \langle \tau(x), \gamma \rangle.
\end{aligned}$$

It follows that $\sigma(x) = \tau(x)$ for all $x \in K^n$, or $\sigma = \tau$.

Clearly, T has integer entries. Moreover, there is an integral matrix S such that $\tau^{-1}(x) = Sx$ modulo one.

It follows that $(TS - I)x$ and $(ST - I)x$ have integer components *for all* $x \in K^n$. But this can only be true if $TS - I = ST - I = 0$ is the zero matrix.

Thus $T^{-1} = S$ also has integer entries. Hence $\det(T)$ and $\det(T^{-1})$ are integers. Finally, $\det(T)\det(T^{-1}) = \det(I) = 1$ implies that $\det(T) = \det(T^{-1}) = \pm 1$. ∎

Example 3 Let G_0 be an arbitrary compact abelian group and define $G = \bigotimes_{n=-\infty}^{\infty} G_n$ to be the countable direct product (complete direct sum) of copies G_n of G_0. Define $\sigma\colon G \to G$ by $\sigma(x) = y$, where $y_n = x_{n-1}$. The transformation σ is a *generalized shift* or *symbolic flow* on the group G_0 and is easily seen to be a bicontinuous group automorphism. In case G_0 is the finite group $G_0 = \{0, 1, \ldots, k-1\}$ of integers modulo k, σ is the inverse of the transformation of Example 3, Chapter I, and Example 1, Chapter II.

The dual $\hat{G}$ of the compact group G is the discrete group $\hat{G} = \bigoplus_{n=-\infty}^{\infty} \hat{G}_n$, where $\hat{G}_n = \hat{G}_0$ is the discrete dual of G_0 and the sum is the algebraic direct sum. Thus a typical element $\gamma \in \hat{G}$ is a sequence $\gamma = (\gamma_n)$, where $\gamma_n = 0$ for all but a finite number of values of n. For $x \in G$, $\gamma \in \hat{G}$, we set $\langle x, \gamma\rangle = \prod_{n=-\infty}^{\infty} \langle x_n, \gamma_n\rangle$. Direct computation shows that the adjoint σ^* of σ is given by $\sigma^*(\gamma) = \delta$, where $\delta_n = \gamma_{n+1}$.

Example 4 Let G and σ be as in the previous example. Define $\tau\colon G \to G$ by $\tau(x) = \sigma(x) + x$. Then $\tau(x + y) = \tau(x) + \tau(y)$, so that τ is a (continuous) endomorphism. It is not invertible since $\tau(x) = 0$, for example, if $x_{2n} = a \neq 0$ and $x_{2n-1} = -a$ for each n. It is, however, an epimorphism and so (Exercise 1) preserves the Haar measure on X. To see that τ is epic, let $y \in G$ be arbitrary and note that $y = \tau(x)$ is equivalent to the infinite system of equations $y_n = x_{n-1} + x_n$. We can, for example, set $x_0 = 0$ and define inductively

$$
\begin{aligned}
x_n &= y_n - x_{n-1} \qquad (n = 1, 2, \ldots)\\
&= y_{n+1} - x_{n+1} \qquad (n = -1, -2, \ldots).
\end{aligned}
$$

The adjoint τ^* of τ is given by $\tau^*(\gamma) = \sigma^*(\gamma) + \gamma$.

Example 5 We construct a slightly different example by taking $G = \bigotimes_{n=1}^{\infty} G_n$ to be a one-sided direct product of copies G_n of G_0. We define σ^* and τ^* on $\hat{G} = \bigoplus_{n=1}^{\infty} G_n$ as above:

$$
\begin{aligned}
\sigma^*(\gamma) &= \delta \qquad \text{where} \quad \delta_n = \gamma_{n+1},\\
\tau^*(\gamma) &= \sigma^*(\gamma) + \gamma.
\end{aligned}
$$

It follows that σ and τ are endomorphisms of G with

$$
\begin{aligned}
\sigma(x) &= y \qquad \text{where} \quad y_1 = 0, \quad y_n = x_{n-1} \quad (n \geq 2),\\
\tau(x) &= \sigma(x) + x.
\end{aligned}
$$

In this case, σ is *not* an epimorphism, since the first component of $\sigma(x)$ is always 0. On the other hand, τ is an *automorphism*, since the system of equations

$$x_1 = y_1$$
$$x_n + x_{n-1} = y_n \qquad (n \geq 2)$$

has a unique solution $x \in G$ for each $y \in G$.

Now let $a_0 \in G_0$ and define $a \in G$ to be $a = (a_0, 0, 0, \ldots)$. The *affine* transformation ϕ defined by

$$\phi(x) = \tau(x) + a = \sigma(x) + x + a$$

will prove to be central in the later development of transformations with quasidiscrete spectrum. Note that an explicit formula for ϕ is given by

$$\phi(x_1, x_2, x_3, \ldots) = (x_1 + a_0, x_2 + x_1, x_3 + x_2, \ldots).$$

Example 6 Let $G_n = \{0, 1, \ldots, k_n - 1\}$ be a finite cyclic group of order k_n $(n = 1, 2, \ldots)$. Define G as a set to be $\times_{n=1}^{\infty} G_n$. Let G have the product topology and make it into a group by setting $x + y = z$, where z_n is defined inductively by

$$x_1 + y_1 = z_1 + w_1 k_1; \qquad 0 \leq z_1 < k_1, \quad w_1 = 0, 1$$
$$x_{n+1} + y_{n+1} + w_n = z_{n+1} + w_{n+1} k_{n+1}; \qquad 0 \leq z_{n+1} < k_{n+1}, \quad w_{n+1} = 0, 1. \tag{5}$$

The idea is that we perform addition component by component with "carry," and the action resembles that of an adding machine with variably sized counters for different digits.

Let $a = (1, 0, 0, \ldots) \in G$. The element a is a *topological generator* of G in the sense that $\{na : n = 0, 1, 2, \ldots\}$ is dense in G. A compact group with a topological generator is said to be *monothetic*. Let ϕ: $G \to G$ be translation by a, that is, $\phi(x) = x + a$. The transformation ϕ is the adding machine transformation of Exercise 5, Chapter I.

Now let $H_n = \times_{k=1}^{n} G_k$ and define addition as in (5) on H_n. (The "carry" w_n is simply dropped.) Then (i) H_n is a cyclic group of order $k_1 k_2 \cdots k_n$, (ii) the natural projection of G onto H_n carries ϕ onto an ergodic translation ϕ_n of the finite group H_n, and (iii) relative to these projections, the system (G, ϕ) is an inverse limit (in the category of affine transformations on compact groups) of the systems (H_n, ϕ_n). (Cf. Exercise 23, Chapter I.)

2. ERGODICITY

Our next project is to find conditions under which the epimorphism τ or the affine map ϕ is ergodic. Throughout this section τ will be a continuous epimorphism of the compact abelian group G and ϕ will be the affine map $\phi(x) = \tau(x) + a$ for some fixed $a \in G$. We will denote by σ the endomorphism $\sigma(x) = \tau(x) - x$.

Note that the adjoint τ^* of the epimorphism τ is a monomorphism of $\hat{G}$. This is true because

$$\begin{aligned}\tau^*(\gamma) = 0 &\Rightarrow \langle x, \tau^*(\gamma)\rangle = \langle \tau(x), \gamma\rangle = 1 \qquad \text{for all} \quad x \in G \\ &\Rightarrow \gamma = 0\end{aligned}$$

since τ is epic. Our first result is that τ is ergodic iff τ^* has no finite orbits in $\hat{G}$ except at 0.

Theorem 3.1 *The epimorphism τ is ergodic iff the orbit $O_{\tau^*}^+(\gamma)$ is infinite for each $\gamma \neq 0$.*

Proof Suppose τ is ergodic and $O_{\tau^*}^+(\gamma)$ is finite; that is, there exist nonnegative integers $n_1 < n_2$ such that

$$O_{\tau^*}^+(\gamma) = \{\gamma, \tau^*(\gamma), \ldots, \tau^{*(n_2-1)}(\gamma)\}, \qquad \tau^{*n_2}(\gamma) = \tau^{*n_1}(\gamma).$$

Let $f \in L_2(G)$ be defined by

$$\begin{aligned}f(x) &= \langle \tau^{n_1}(x), \gamma\rangle + \cdots + \langle \tau^{n_2-1}(x), \gamma\rangle \\ &= \langle x, \tau^{*n_1}(\gamma)\rangle + \cdots + \langle x, \tau^{*(n_2-1)}(\gamma)\rangle.\end{aligned}$$

Then

$$\begin{aligned}f(\tau(x)) &= \langle x, \tau^{*(n_1+1)}(\gamma)\rangle + \cdots + \langle x, \tau^{*n_2}(\gamma)\rangle \\ &= f(x)\end{aligned}$$

for all $x \in G$. It follows that f must be constantly equal to $f(0) = n_2 - n_1 \neq 0$. Since τ is measure preserving, we have

$$n_2 - n_1 = f(0) = \int f\, dm = (n_2 - n_1) \int \gamma\, dm,$$

$$\int \gamma\, dm = 1.$$

But this means that $\gamma = 0$, as required.

Conversely, suppose that $O^+_{\tau^*}(\gamma)$ is infinite for $\gamma \neq 0$, and let $f \in L_2(G)$ be τ-invariant. Then for each $\gamma \in \hat{G}$ we have

$$\hat{f}(\tau^*(\gamma)) = \int f(x)\langle \overline{\tau(x), \gamma}\rangle m(dx) = \int f(\tau(x))\langle \overline{\tau(x), \gamma}\rangle\, m(dx)$$

$$= \int f(x)\langle \overline{x, \gamma}\rangle m(dx) = \hat{f}(\gamma).$$

That is, $\hat{f}$ is τ^*-invariant. For $\gamma_0 \neq 0$ the points in $O^+_{\tau^*}(\gamma_0)$ are distinct, and so

$$\sum_{n=0}^{\infty} |\hat{f}(\tau^{*n}(\gamma_0))|^2 \leq \sum_{\gamma \in \hat{G}} |\hat{f}(\gamma)|^2 < \infty,$$

since $\hat{f} \in L_2(\hat{G})$. But each term in the sum on the left is equal to $|\hat{f}(\gamma_0)|^2$, hence must be zero. Since $\hat{f}(\gamma) = 0$ except for $\gamma = 0$, $\hat{f}$ is a constant. Therefore, τ is ergodic. ∎

Remark If τ is an automorphism, then $O^+_{\tau^*}(\gamma)$ is finite iff $\tau^{*k}(\gamma) = \gamma$ for some $k \geq 1$.

Corollary 3.1.1 *If τ is ergodic, then it is strongly mixing.*

Proof Suppose τ is ergodic and $\gamma_1, \gamma_2 \in \hat{G}$ with not both of γ_1, γ_2 equal to 0. Then $\tau^{*n}(\gamma_1) = \gamma_2$ for at most one value of n. Therefore,

$$(T_\tau^n \gamma_1, \gamma_2) = \int (\tau^{*n}\gamma_1)\bar{\gamma}_2\, dm = 0$$

for all sufficiently large n. Hence, if f and g are finite linear combinations of functions in $\hat{G}$, then

$$(T_\tau^n f, g) = (f, 1)(1, g)$$

for all sufficiently large n. Given any $f, g \in L_2(G)$ and $\varepsilon > 0$, choose $f_\varepsilon, g_\varepsilon$ in the linear span of $\hat{G} \subseteq L_2(G)$ such that $\|f - f_\varepsilon\|_2 < \varepsilon$, $\|g - g_\varepsilon\|_2 < \varepsilon$. Then there exists n_0 such that

$$\begin{aligned} n \geq n_0 \Rightarrow |(T_\tau^n f, g) - (f, 1)(1, g)| &\leq |(T_\tau^n[f - f_\varepsilon], g)| \\ &+ |(T_\tau^n f_\varepsilon, g - g_\varepsilon)| + |(f_\varepsilon, 1) - (f, 1)|\,|(1, g_\varepsilon)| \\ &+ |(f, 1)|\,|(1, g_\varepsilon) - (1, g)| \\ &< \varepsilon\|g\| + \varepsilon\|f_\varepsilon\| + \varepsilon\|g_\varepsilon\| + \varepsilon\|f\| < \varepsilon(2\|g\| + 2\|f\| + 2\varepsilon). \end{aligned}$$

Hence

$$\lim_{n\to\infty} (T_\tau^n f, g) = (f, 1)(1, g).$$

In particular, setting $f = \chi_A$, $g = \chi_B$ yields

$$\lim_{n\to\infty} m(A \cap \tau^{-n}(B)) = m(A)\, m(B),$$

so that τ is strongly mixing. ▮

Corollary 3.1.2 *A continuous automorphism τ of the l torus K^l (Example 2) is ergodic iff the associated unimodular matrix $T = (t_{ij})$ has no eigenvalue which is a root of unity.*

Proof For each $\gamma \in Z^l$ we have (Exercise 2) that $\tau^*(\gamma) = \gamma T$. Suppose that τ is not ergodic, and hence that $O_{\tau^*}^+(\gamma) = \{\gamma, \tau^*(\gamma), \ldots, \tau^{*(n-1)}(\gamma)\}$ for some $\gamma \neq 0$ and some positive integer n. Thus $\gamma T^n = \gamma$. Let λ be any primitive nth root of unity, and define the l-dimensional vector $f = (f_1, \ldots, f_l)$ by

$$f = \lambda^{n-1}\gamma + \lambda^{n-2}(\gamma T) + \cdots + (\gamma T^{n-1}). \tag{6}$$

Then

$$fT = \lambda^{n-1}(\gamma T) + \cdots + \lambda(\gamma T^{n-1}) + \gamma = \lambda f.$$

It only remains to show that f is not the zero vector.

Suppose $f = 0$. Since $\gamma \neq 0$, (6) represents l polynomial equations of degree no greater than $n - 1$ in λ and having integer coefficients. But the only such equation is

$$\lambda^{n-1} + \lambda^{n-2} + \cdots + 1 = 0.$$

This contradicts the assumption that $\gamma, \gamma T, \ldots, \gamma T^{n-1}$ are distinct.

Conversely, suppose that $\lambda^n = 1$ and λ is an eigenvalue of the integral matrix T. It follows that $T^n - I$ is singular. This means that there exists a nonzero $f \in R^l$ with rational components such that $fT^n = f$. It follows that there exists such an f with integer components, that is, $f \in Z^l$. But then $\tau^{*n}(f) = fT^n = f$, $O_{\tau^*}^+(f)$ is finite, and τ is not ergodic. ▮

Examples There are no ergodic automorphisms of the one-torus or circle group K. On the two-torus K^2 there are many ergodic automorphisms. Consider the unimodular 2×2 matrix $T = (t_{ij})$. Let t denote the trace $t = t_{11} + t_{22}$ of T. Thus the characteristic equation of T is

$$\lambda^2 - t\lambda \pm 1 = 0.$$

If $t = 0$, then the eigenvalues are either ± 1 or $\pm i$. In either case, τ is nonergodic. In general, the roots λ_1 and λ_2 must be either real or complex conjugates satisfying $\lambda_1 \lambda_2 = \pm 1$.

Case 1 $\det(T) = -1$. The characteristic equation is

$$\lambda^2 - t\lambda - 1 = 0,$$

and there are no complex roots. For $t \neq 0$, the roots are real and distinct with $|\lambda_1| < 1$ and $|\lambda_2| > 1$. Thus τ is ergodic.

Case 2 $\det(T) = 1$, $t^2 > 4$. In this case also, the roots are real and distinct, and τ is ergodic.

Case 3 $\det(T) = 1, 0 < t^2 \leq 4$. Since t is an integer, the only possibilities are $t = \pm 1, \pm 2$. The corresponding pairs of roots in the four cases are

$$\frac{1}{2} \pm \frac{\sqrt{3}}{2} i; \qquad -\frac{1}{2} \pm \frac{\sqrt{3}}{2} i; \qquad 1, 1; \qquad -1, -1.$$

Each is a root of unity, and so τ is nonergodic. We have proved the following proposition.

Proposition 3.2 *The automorphism τ of the two-torus is ergodic iff the corresponding unimodular matrix T has* (i) *determinant* 1 *and nonzero trace or* (ii) *determinant* -1 *and trace greater than* 2 *in absolute value.*

The epimorphisms σ and τ of Examples 3 and 4 are ergodic (Exercise 9). The automorphism τ of Example 5 is *not* ergodic. To see this, let $\gamma = (\xi, 0, 0, 0, \ldots)$, where $\xi \in \hat{G}_0$, $\xi \neq 0$. Then $\tau^*(\gamma) = \gamma$.

Ergodicity of the epimorphism τ has interesting and immediate implications regarding the affine transformation $\phi(x) = \tau(x) + a$, which we state in the following theorem.

Theorem 3.2 *Let τ be a continuous automorphism of G, and let $\phi(x) = \tau(x) + a$, where $a \in G$. Then the following are equivalent:*

(i) *τ^* has no finite orbits in $\hat{G} \sim \{0\}$.*
(ii) *τ is ergodic.*
(iii) *τ is strongly mixing.*
(iv) *ϕ is strongly mixing.*
(v) *ϕ is weakly mixing.*

Proof We have already shown that (i), (ii), and (iii) are equivalent. To show that (i) implies (iv), note that

$$\phi^n(x) = \tau^n(x) + \sum_{j=0}^{n-1} \tau^j(a)$$
$$\gamma(\phi^n(x)) = \gamma(\tau^n(x)) \prod_{j=0}^{n-1} \gamma(\tau^j(a)) \tag{7}$$

so that $T_\phi{}^n\gamma$ is a constant multiple (varying with n) of $\tau^{*n}(\gamma)$ for each $\gamma \in \hat{G}$. It follows as in the proof of Corollary 3.1.1 that $(T_\phi{}^n\gamma_1, \gamma_2) = 0$ for sufficiently large n whenever γ_1 and γ_2 are not both the zero character. The rest of the proof is unchanged.

Since (iv) always implies (v), it only remains to show that (v) implies (i).

Suppose (i) is false, say $\tau^{*n}(\gamma) = \gamma$, $\gamma \neq 0$, with $n \geq 1$ minimal. We can then write (7) as

$$T_\phi{}^n\gamma(x) = \lambda(\gamma)\,\gamma(x)$$

where $|\lambda(\gamma)| = 1$. If M is the linear span in $L_2(G)$ of the set $\{\gamma, T_\phi \gamma, \ldots, T_\phi^{n-1}\gamma\}$, it follows that $T_\phi(M) \subseteq M$. Moreover, $1 \notin M$ since the characters $0, \gamma, \tau^*(\gamma), \ldots, \tau^{*(n-1)}(\gamma)$ are assumed distinct and hence linearly independent in $L_2(G)$. [Note that the character 0 is the constant function $1 \in L_2(G)$.] Since M is finite-dimensional, T_ϕ must have an eigenfunction $g \in M$. According to Theorem 1.6, this means that ϕ is not weakly mixing. ∎

Note that the above theorem does not include ergodicity of ϕ as one of the equivalent conditions. Indeed, ϕ may be ergodic (but not mixing) even when τ is nonergodic. For example, translation by an irrational number, $\phi(x) = x + a$, on the circle group is ergodic, but $\tau(x) = x$ is not. On the other hand, notice that $\phi(x) = -x + a$ is never ergodic since ϕ^2 is the identity.

A measure-preserving transformation ϕ is said to be *totally ergodic* if ϕ^n is ergodic for each $n \geq 1$.

Theorem 3.3 *Let τ be a continuous automorphism of the compact abelian group G, and assume that $\hat{G}$ has no nonzero elements of finite order. Let ϕ and σ be as before. Then the following are equivalent:*

(i) *ϕ is ergodic.*
(ii) *ϕ is totally ergodic.*
(iii) $\tau^{*n}(\gamma) = \gamma$, $\lambda_n(\gamma) = \prod_{k=0}^{n-1} \gamma(\tau^k(a)) = 1 \Rightarrow \gamma = 0$.

(iv) $\tau^{*n}(\gamma) = \gamma \Rightarrow \tau^*(\gamma) = \gamma$, *and* $\tau^*(\gamma) = \gamma$, $\gamma(a) = 1 \Rightarrow \gamma = 0$.

(v) $\tau^{*n}(\gamma) = \gamma \Rightarrow \tau^*(\gamma) = \gamma$, *and the group* $G(a, \sigma) = \{na + \sigma(x) : n \in Z, x \in G\}$ *is dense in* G.

In any case, (iii) *implies* (ii) *and* (iv) *implies* (i).

Proof We have shown [Eq. (7)] that

$$T_\phi{}^n \gamma = \lambda_n(\gamma)\tau^{*n}(\gamma)$$

for each $n \geq 1$ and each $\gamma \in \hat{G}$. Suppose now that ϕ is ergodic, that $\tau^{*n}(\gamma) = \gamma$, and that $\lambda_n(\gamma) = 1$. Let l be the smallest positive integer such that $\tau^{*l}(\gamma) = \gamma$. Then $\gamma, \tau^*(\gamma), \ldots, \tau^{*(l-1)}(\gamma)$ are distinct characters, and l divides n, say $n = jl$. Moreover, $\tau^{*(il+k)}(\gamma) = \tau^{*k}(\gamma)$ for all positive integers i and k. Let us show that $[\lambda_{il+k}(\gamma)]^n = [\lambda_k(\gamma)]^n$. Since $\tau^{*l}(\gamma) = \gamma$,

$$\lambda_{il+k}(\gamma) = \prod_{t=0}^{il+k-1} \gamma(\tau^t a) = \left[\prod_{t=0}^{l-1} \gamma(\tau^t a)\right]^i \prod_{t=0}^{k-1} \gamma(\tau^t a) = [\lambda_l(\gamma)]^i \lambda_k(\gamma).$$

In particular,

$$\lambda_n(\gamma) = [\lambda_l(\gamma)]^j = 1.$$

Thus

$$[\lambda_{il+k}(\gamma)]^n = [\lambda_l(\gamma)]^{ijl}[\lambda_k(\gamma)]^n = [\lambda_k(\gamma)]^n.$$

Now define $f \in L_2(G)$ by

$$\begin{aligned} f(x) &= \sum_{k=0}^{n-1} [T_\phi{}^k \gamma(x)]^n = \sum_{i=0}^{j-1} \sum_{k=0}^{l-1} [T_\phi^{il+k} \gamma(x)]^n \\ &= \sum_{i=0}^{j-1} \sum_{k=0}^{l-1} [\lambda_{il+k}(\gamma)]^n [\tau^{*(il+k)} \gamma(x)]^n \\ &= \sum_{k=0}^{l-1} j[\lambda_k(\gamma)]^n \, \tau^{*k}(n\gamma)(x). \end{aligned} \tag{8}$$

Since $\hat{G}$ has no elements of finite order other than 0, and since $\gamma, \tau^*\gamma, \ldots, \tau^{*(l-1)}\gamma$ are distinct characters, it follows that $n\gamma, \tau^*(n\gamma), \ldots, \tau^{*(l-1)}(n\gamma)$ are distinct characters. On the other hand, since $\lambda_n(\gamma) = 1$, we have $T_\phi{}^n \gamma = \gamma$, and so $T_\phi f = f$. By the ergodicity of ϕ, it follows that f is a constant function. Because of the linear independence of distinct characters, this can only happen if $l = 1$ and $\gamma = 0$. Thus (i) implies (iii).

(iii) $\Rightarrow$ (iv). Suppose $\tau^{*n}(\gamma) = \gamma$, and let $\xi = \sigma^*(\gamma)$. Since σ commutes with τ, we have

$$\tau^{*n}(\xi) = \tau^{*n}\sigma^*(\gamma) = \sigma^*\tau^{*n}(\gamma) = \sigma^*(\gamma) = \xi.$$

Moreover,

$$\lambda_n(\xi) = \prod_{k=0}^{n-1} \tau^{*k}\xi(a) = \prod_{k=0}^{n-1} \sigma^*\tau^{*k}\gamma(a)$$
$$= \prod_{k=0}^{n-1} \tau^{*(k+1)}\gamma(a)/\tau^{*k}\gamma(a) = \tau^{*n}\gamma(a)/\gamma(a) = 1.$$

According to (iii), we have $\xi = 0$, that is, $\tau^*(\gamma) = \gamma$. Since $\lambda_1(\gamma) = \gamma(a)$, the second statement in (iv) is a special case of (iii), and we have completed the proof of (iii) $\Rightarrow$ (iv).

(iv) $\Rightarrow$ (ii). Suppose $f \in L_2(G)$, $T_\phi{}^n f = f$. Then f can be represented by a Fourier series

$$f = \sum_{\gamma \in \hat{G}} (f, \gamma)\gamma, \qquad \text{and} \qquad T_\phi{}^n f = \sum_{\gamma \in \hat{G}} (f, \gamma)\, \lambda_n(\gamma)\, \tau^{*n}\gamma.$$

Equating coefficients gives

$$(f, \tau^{*n}\gamma) = \lambda_n(\gamma)\,(f, \gamma)$$

for each $\gamma \in \hat{G}$. In particular,

$$|(f, \tau^{*n}\gamma)| = |(f, \gamma)| \qquad (\gamma \in \hat{G})$$

and

$$|(f, \tau^{*jn}\gamma)| = |(f, \gamma)| \qquad (\gamma \in \hat{G}, j \geq 1).$$

Since $\sum_{\gamma \in \hat{G}} |(f, \gamma)|^2 < \infty$, we must have $(f, \gamma) = 0$ for each γ with infinite orbit. But by (iv), each γ with finite orbit satisfies $\tau^*\gamma = \gamma$, and hence $T_\phi\, \gamma = \gamma(a)\gamma$. Thus

$$f = \sum_{\tau^*\gamma = \gamma} (f, \gamma)\gamma \tag{9}$$

and

$$T_\phi{}^n f = \sum_{\tau^*\gamma = \gamma} (f, \gamma)[\gamma(a)]^n\gamma. \tag{10}$$

Since by (iv)

$$\tau^*\gamma = \gamma \Rightarrow \tau^*(n\gamma) = n\gamma \Rightarrow (n\gamma)(a) = [\gamma(a)]^n \neq 1$$

except when $\gamma = 0$, it follows by equating coefficients in (9) and (10) that $(f, \gamma) = 0$ for all $\gamma \neq 0$. That is, f is a constant. Thus T_ϕ is totally ergodic.

Since (ii) clearly implies (i), the proof of the theorem will be complete if we can show that

$$\tau^*\gamma = \gamma,\ \gamma(a) = 1 \Rightarrow \gamma = 0$$

iff the group generated by a and $\sigma(G)$ is dense in G. But this latter condition is equivalent to

$$(a)^{\perp} \cap \sigma(G)^{\perp} = 0,$$

where (a) denotes the cyclic group generated by a and $\perp$ indicates the annihilator. Now

$$\gamma \in (a)^{\perp} \quad \text{iff} \quad \gamma(a) = 1$$

and

$$\begin{aligned} \gamma \in \sigma(G)^{\perp} \quad & \text{iff} \quad \langle \tau x - x, \gamma \rangle = 1 \quad (x \in G) \\ & \text{iff} \quad \langle \tau x, \gamma \rangle = \langle x, \gamma \rangle \quad (x \in G) \\ & \text{iff} \quad \tau^* \gamma = \gamma. \end{aligned}$$

This completes the proof when $\hat{G}$ is torsion-free. The last statement of the theorem is proved similarly (Exercise 13). ∎

Definition 3.1 The affine transformation ϕ is said to be *semiergodic* if one of the following three equivalent conditions is satisfied:

$$\tau^* \gamma = \gamma,\ \gamma(a) = 1 \Rightarrow \gamma = 0 \tag{11}$$

$$[(a) + \sigma(G)]^{-} = G \tag{12}$$

$$(a)^{\perp} \cap \sigma(G)^{\perp} = (0). \tag{13}$$

Example 5a Let ϕ be defined as in Example 5. Suppose that $\tau^* \gamma = \gamma$, that is, $\sigma^* \gamma = 0$. Then $\gamma = (\gamma_1, 0, 0, \ldots)$. If $a_0 \in G_0$ separates the points of $\hat{G}_0$, that is, $\langle a_0, \gamma_0 \rangle = 1$ only for $\gamma_0 = 0$, then ϕ is *semiergodic*. In particular, this is true if a_0 is a topological generator of G_0.

If G_0 has no nonzero elements of finite order, then G has no nonzero elements of finite order. In this case,

$$\begin{aligned} \tau^{*k}(\gamma) = \gamma \Rightarrow & \sum_{j=1}^{k} \binom{k}{j} \gamma_{n+j} = 0 \qquad (n = 1, 2, \ldots) \\ \Rightarrow & \tau^*(\gamma) = \gamma. \end{aligned}$$

For if $\gamma \neq 0$, let p be the largest integer with $\gamma_p \neq 0$, and assume that $p > 1$. Setting $n = p - 1$, we see that

$$\sum_{j=1}^{k} \binom{k}{j} \gamma_{n+j} = k\gamma_p = 0.$$

Since G_0 is torsion-free, $\gamma_p = 0$, a contradiction.

Summarizing, if G_0 has no nonzero elements of finite order and if a_0 separates the points of $\hat{G}_0$, then (iv) is satisfied and ϕ is *ergodic*.

Example 5b Let ϕ be defined as in Example 5, and let $G_0 = \hat{G}_0 = \{0, 1, \ldots, 15\}$ with addition modulo 16 as the group operation and

$$\langle x, \gamma \rangle = e^{ix \cdot \gamma/8}.$$

Then $a_0 = 1$ is a generator for G_0, and so ϕ is semiergodic. Let us show that ϕ is *not* ergodic.

Consider

$$\gamma = (5, 6, 4, 8, 0, 0, \ldots).$$

Then

$$\tau^* \gamma = (11, 10, 12, 8, 0, 0, \ldots)$$
$$\tau^{*2} \gamma = (5, 6, 4, 8, 0, 0, \ldots) = \gamma,$$

so that (iv) is not satisfied. Moreover, $f(x) = [\gamma(x)]^2 + [T_\phi \gamma(x)]^2$ is a nontrivial invariant function, and so ϕ is not ergodic.

3. DISCRETE AND QUASIDISCRETE SPECTRUM

A measure-preserving transformation ϕ has *discrete spectrum* if the eigenfunctions of the associated operator T_ϕ on L_2 form a basis for L_2. A classical result of Halmos and von Neumann [34] asserts that the ergodic measure-preserving transformation ϕ has discrete spectrum iff it is measure-theoretically isomorphic to translation by a topological generator on a monothetic group.

In 1962, Abramov [2], using a notion of quasieigenfunction introduced by Halmos [32], gave a definition of *quasidiscrete spectrum* and generalized the theorem of Halmos and von Neumann. The appropriate model is an ergodic affine transformation satisfying an additional condition to be discussed presently.

In 1965, Hahn and Parry [29] defined quasidiscrete spectrum for a minimal homeomorphism of a compact Hausdorff space and proved the corresponding isomorphism theorem. In 1969, Brown [11] showed that transformations with quasidiscrete spectrum in both the topological and measure-theoretic cases are factors of a single affine transformation defined in terms of the shift on a certain product group.

Let $\phi(x) = \tau(x) + a$ be an affine transformation of the compact abelian group G, where τ is assumed to be an automorphism. In order to describe the condition referred to above, we first define a sequence of subgroups of the dual group $\hat{G}$.

Let Γ_0 be the trivial group consisting only of the identity of $\hat{G}$,

$$\Gamma_0 = \{0\},$$

and define Γ_n inductively by

$$\Gamma_{n+1} = \{\gamma \in \hat{G} : \tau^*\gamma - \gamma \in \Gamma_n\}.$$

Finally, let

$$\Gamma = \bigcup_{n=0}^{\infty} \Gamma_n.$$

Note that Γ_n is the kernel of the group endomorphism σ^{*n} and that $\Gamma_n \subseteq \Gamma_{n+1}$ for each n. Therefore each Γ_n and also Γ are groups. Moreover, it follows easily that $\tau^*(\Gamma_n) = \Gamma_n$ and $\tau^*(\Gamma) = \Gamma$.

Definition 3.2 The affine transformation ϕ has *quasidiscrete spectrum* if $\Gamma = \hat{G}$.

Note that $\Gamma = G_0$ when τ is ergodic. Thus the transformations discussed in this section lie at the opposite end of the spectrum among affine transformations from those for which τ is ergodic.

Note also that ϕ has quasidiscrete spectrum iff $\bigcap_{n=0}^{\infty} \sigma^n(G) = \{0\}$ (Exercise 21).

Examples Of those in Section 1 only Examples 5 and 6 have quasi-discrete spectrum. Let us show this for the ϕ of Example 5. We have

$$\tau^*(\gamma_1, \gamma_2, \ldots) = (\gamma_1 + \gamma_2, \gamma_2 + \gamma_3, \ldots)$$

so that $\tau^*\gamma = \gamma$ iff $\gamma_2 = \gamma_3 = \cdots = 0$. That is,

$$\Gamma_1 = \{\gamma : \gamma_k = 0 \ (k \neq 1)\}.$$

Similarly, it is easily proved inductively that

$$\Gamma_n = \{\gamma : \gamma_k = 0 \ (k > n)\}.$$

Thus

$$\Gamma = \bigcup_{n=0}^{\infty} \Gamma_n = \bigoplus_{n=1}^{\infty} \hat{G}_0 = \hat{G}.$$

Note that $\hat{G} = \Gamma$ is the inductive limit of the sequence of groups Γ_n, and hence G is an inverse limit of the sequence $G_n = \hat{\Gamma}_n$,

$$G_n \cong G/\Gamma_n^{\perp} \cong \bigotimes_{k=1}^{n} G_0.$$

It follows easily that $\Phi = (G, \phi)$ is affinely isomorphic to the inverse limit of the sequence $\Phi_n = (G_n, \phi_n)$, where

$$\phi_n(x_1, x_2, \ldots, x_n) = (x_1 + a_0, x_1 + x_2, \ldots, x_{n-1} + x_n).$$

The fact that the ϕ of Example 6 has quasidiscrete spectrum will follow from the following proposition.

Proposition 3.3 *Let a be a topological generator of the monothetic group G, and let $\phi(x) = x + a$. Then $\Gamma_1 = \hat{G}$, so that ϕ has quasidiscrete spectrum.*

Indeed, ϕ has discrete spectrum, since $\hat{G}$ is a basis for $L_2(G)$. The proof of this proposition is left to the exercises.

Let us turn now to a discussion of quasidiscrete spectrum for abstract and classical dynamical systems. Suppose that $\Phi = (X, \mathscr{B}, \mu, \phi)$ is an abstract dynamical system. We shall assume that Φ is ergodic. Recall that $f \in L_2(X)$ is an eigenfunction for T_ϕ with eigenvalue $\lambda \in K$ if $T_\phi f = \lambda f$. Let us denote by A_1 the set of eigenvalues of T_ϕ and by B_1 the set of eigenfunctions having constant absolute value one. According to Theorem 1.6, A_1 is a group, $1 \in A_1 \subseteq K \subseteq B_1$, and any two elements f, $g \in B_1$ associated with the same eigenvalue λ satisfy $f/g \in K$. This last assertion must, of course, be interpreted to mean that f/g is almost everywhere equal to a constant, and the absolute value of that constant is one.

For completeness let us set $A_0 = \{1\}$ and $B_0 = K$. We shall then define A_n, $B_n \subseteq L_2(X)$ inductively by

$$g \in A_{n+1}, \quad f \in B_{n+1} \qquad \text{iff} \quad g \in B_n, \quad |f| = 1 \quad \text{and} \quad T_\phi f = gf,$$

and set $A = \bigcup_{n=0}^{\infty} A_n$, $B = \bigcup_{n=0}^{\infty} B_n$. The elements of A are called *quasi-eigenvalues* of Φ, the elements of B *quasieigenfunctions*.

Definition 3.3 Φ has *quasidiscrete spectrum* if the linear span of B is dense in $L_2(X)$.

Proposition 3.4 *For each $n = 0, 1, 2, \ldots$, A_n and B_n are groups (pointwise almost everywhere multiplication), $A_n \subseteq A_{n+1} \subseteq B_n \subseteq B_{n+1}$, and A_{n+1}/A_n is isomorphic to a subgroup of K.*

Proof Let $\mathscr{G} = \mathscr{G}(X, \mathscr{B})$ be the group of complex-valued, $\mathscr{B}$-measurable functions f on X with $|f(x)| \equiv 1$. We define homomorphisms $\tilde{\tau}$ and $\tilde{\sigma}$ on $\mathscr{G}$ by

$$\tilde{\tau} f(x) = T_\phi f(x) = f(\phi(x)) \tag{14}$$

$$\tilde{\sigma} f(x) = T_\phi f(x)/f(x). \tag{15}$$

By ergodicity of ϕ, the kernel of $\tilde{\sigma}$ is $K = B_0$. It follows easily from (15) that $B_{n+1} = \tilde{\sigma}^{-1}(B_n)$, and hence that B_n $(n = 0, 1, 2, \ldots)$ is the kernel of $\tilde{\sigma}^{(n+1)}$. Moreover,

$$A_n = \tilde{\sigma}(B_n) \subseteq B_{n-1} \qquad (n = 1, 2, \ldots).$$

Hence each of the A_n and B_n are subgroups of $\mathscr{G}$. Clearly, the kernel of $\tilde{\sigma}^{n+1}$ is contained in the kernel of $\tilde{\sigma}^{n+2}$, so that $B_n \subseteq B_{n+1}$, and $A_n = \tilde{\sigma}(B_n) \subseteq \tilde{\sigma}(B_{n+1}) = A_{n+1}$.

To prove the last assertion, note that $\tilde{\sigma}^{n+1}(B_{n+1}) \subseteq K$ and recall that the kernel of $\tilde{\sigma}^{n+1}$ is B_n. It follows that $\tilde{\sigma}^n$ maps $A_{n+1} = \tilde{\sigma}(B_{n+1})$ into K, and the kernel of $\tilde{\sigma}^n$ restricted to A_{n+1} is A_n. For if $g = \tilde{\sigma}(f) \in A_{n+1}$, with $f \in B_{n+1}$, and if $\tilde{\sigma}^n(g) = \tilde{\sigma}^{n+1}(f) = 1$, then $f \in B_n$, and so $g \in A_n$. On the other hand, $A_n \subseteq B_{n-1}$, which is the kernel of $\tilde{\sigma}^n$. Thus

$$A_{n+1}/A_n \cong \tilde{\sigma}^n(A_{n+1}) \subseteq K. \blacksquare$$

Previously, we have been using the symbol K to denote the compact topological group of complex numbers with modulus one. On the other hand, the preceding proposition is purely group-theoretic in nature since we have as yet imposed no topology on the groups A and B. In the sequel we shall want to consider A to be a discrete topological group. In this context, the last statement of Proposition 3.4 is only true if we replace K by the discrete topological group of complex numbers with modulus one. In order to avoid confusion, we shall henceforth denote this topological group by K_{d}.

The next result shows how we can embed the group A of *quasieigenvalues* of T_ϕ into the countably infinite direct sum $K_{\mathrm{d}}^{\infty} = \bigoplus_{n=1}^{\infty} K_{\mathrm{d}}$ of copies of K_{d}. This result is suggested by the last assertion of Proposition 3.4. The embedding will be accomplished in such a manner that A_n appears as a subgroup of $K_{\mathrm{d}}^{n} \cong \{t \in K_{\mathrm{d}}^{\infty} : t_k = 1 \ (k > n)\}$, the discretized n-torus.

Before proceeding to this result, let us quickly survey the analogous situation for classical dynamical systems. Let $\Phi = (X, \phi)$ be a *minimal* classical dynamical system. Then the only *continuous* invariant functions for T_ϕ are constants. We define subgroups A_n, B_n, A, B of $C(X)$ just as before. Because of minimality, the homomorphism $\tilde{\sigma}$ again has kernel K. Everything else goes through exactly as before. In particular, Proposition 3.4 is valid for Φ.

Definition 3.4 The classical dynamical system $\Phi = (X, \phi)$ has *quasi-discrete spectrum* if the algebra generated by B is dense in $C(X)$.

The group $K_d{}^\infty$ has a natural endomorphism $\hat{\sigma}^*: K_d{}^\infty \to K_d{}^\infty$, called the shift and defined by $\hat{\sigma}^*(t) = u$, where $u_n = t_{n+1}$. This, of course, is a special case of Example 5. The subgroup $\Gamma \subseteq K_d{}^\infty$ is said to be *shift-invariant* if $\hat{\sigma}^*(\Gamma) \subseteq \Gamma$.

Theorem 3.4 *If Φ is either a minimal classical dynamical system or an ergodic abstract dynamical system, then the group A of quasieigenvalues of Φ is isomorphic to a shift-invariant subgroup of $K_d{}^\infty$.*

Proof We have $K = B_0 \subseteq \mathscr{G}$. Since K is a divisible group, it follows [38, p. 11] that the identity map of K into K has an extension to $\mathscr{G}$. That is, there exists a homomorphism $\alpha: \mathscr{G} \to K$ and $\alpha(f) = f$ for $f \in K$. Define $\rho^*: A \to K_d{}^\infty$ by

$$\rho^*(f) = (\alpha(f), \alpha(\tilde{\sigma} f), \alpha(\tilde{\sigma}^2 f), \ldots). \tag{16}$$

Clearly, ρ^* is a homomorphism and $\rho^*\tilde{\sigma} = \hat{\sigma}^*\rho^*$. Since $\tilde{\sigma}(A) \subseteq A$, it follows that $\rho^*(A)$ is a shift-invariant subgroup of $K_d{}^\infty$. It only remains to show that ρ^* is monic.

Suppose then that $f \in A$, $f \neq 1$. It follows that for some $n \geq 0$ we have $f \notin A_n$ and $f \in A_{n+1}$. Hence $\tilde{\sigma}^n f$ is a constant different from 1. Since $\alpha(\tilde{\sigma}^n f) = \tilde{\sigma}^n f$, $\rho^*(f)$ is not the identity. That is, the kernel of ρ^* is trivial. ∎

Theorem 3.5 *If $\Phi = (G, \phi)$ is a semiergodic affine system with quasi-discrete spectrum, then Φ is an algebraic factor of the system $\hat{\Phi} = (\hat{K}_d{}^\omega, \hat{\phi})$ defined as in Example 5, with $\hat{\phi}(x) = \hat{\sigma}(x) + x + \hat{a}$, $\hat{a} = (a_0, 0, 0, \ldots)$, and $a_0(t) = t$.*

Proof First of all, we show that the analog of Theorem 3.4 holds with A replaced by $\Gamma = \hat{G}$. We define $\rho^*: \Gamma \to K_d{}^\infty$ by

$$\rho^*(\gamma) = (\gamma(a), \sigma^*\gamma(a), \sigma^{*2}\gamma(a), \ldots),$$

so that $\rho^*\sigma^* = \hat{\sigma}^*\rho^*$. To show that ρ^* is monic, suppose that $\rho^*(\gamma) = 1$, $\gamma \in \Gamma_{n+1}$, and let $\xi = \sigma^{*n}\gamma$. Then

$$\tau^*\xi = \sigma^*\xi + \xi = \sigma^{*(n+1)}\gamma + \sigma^{*n}\gamma = \sigma^{*n}\gamma = \xi,$$

and $\xi(a) = \sigma^{*n}\gamma(a) = 1$. By semiergodicity, $\xi = 0$. In other words, $\gamma \in \Gamma_n$. By a repetition of this argument, we deduce that $\gamma = 0$.

Now consider the adjoint mapping $\rho: \hat{K}_d{}^\omega \to G = \hat{\Gamma}$. Since ρ^* is monic, ρ is epic. Moreover, $\sigma\rho = \rho\hat{\sigma}$ and hence $\tau\rho = \rho\hat{\tau}$. Noting that

$$\langle \rho(\hat{a}), \gamma \rangle = \langle \hat{a}, \rho^*(\gamma) \rangle = \gamma(a)$$

for each $\gamma \in \Gamma$ implies $\rho(\hat{a}) = a$, we have

$$\rho\hat{\phi}(x) = \rho\hat{\tau}(x) + \rho(\hat{a}) = \tau\rho(x) + a = \phi\rho(x).$$

Thus ρ is a homomorphism of $\hat{\Phi}$ onto Φ as asserted. ∎

In Section 6 we shall show that Theorem 3.5 applies to totally ergodic abstract dynamical systems (Abramov) and to totally minimal classical systems (Hahn–Parry).

4. QUASIPERIODIC SPECTRUM AND THE ERGODIC PART OF τ

In this section we introduce a construction due to Seethoff [56], which permits us to identify the maximal subgroup H of G on which an arbitrary continuous epimorphism τ of G is ergodic. The beginning point is the observation that τ is ergodic on all of G iff τ^* has no nontrivial periodic points. In the extreme opposite case, where H is the zero group, there is a close relationship to the affine transformations with quasidiscrete spectrum, and we say that τ has quasiperiodic spectrum. We show also that such a τ is distal.

Let τ be a continuous epimorphism of G. As in the previous section, we define an increasing sequence of subgroups of $\hat{G}$ as follows:

$$\begin{aligned} \Delta_0 &= \{0\} \\ \Delta_{n+1} &= \{\gamma \in \hat{G} : \tau^{*k}\gamma - \gamma \in \Delta_n \text{ for some positive integer } k\}. \end{aligned} \tag{17}$$

Let $\Delta = \Delta(\tau^*) = \bigcup_{n=1}^{\infty} \Delta_n$.

Note that $\Gamma_n \subseteq \Delta_n$ for each n, and so $\Gamma \subseteq \Delta$.

Definition 3.5 The affine transformation ϕ, defined by $\phi(x) = \tau(x) + a$, has *quasiperiodic spectrum* if $\Delta(\tau^*) = \hat{G}$.

Proposition 3.5 *The epimorphism τ is ergodic iff $\Delta = \Delta_1 = \{0\}$.*

Proof If τ is ergodic, then according to Theorem 3.1, $\Delta_1 = \{0\}$. If τ is not ergodic, there exists a nonzero $\gamma \in \hat{G}$ and positive integers k_1, k_2 such that $\xi = \tau^{*k_1}\gamma = \tau^{*k_1+k_2}\gamma$. But then $\tau^{*k_2}\xi = \xi$, and $\xi \in \Delta_1$. Since τ^* is monic, $\xi \neq 0$.

If $\Delta_1 = \{0\}$, it follows immediately from (17) that $\Delta_n = \{0\}$ for each n, and hence $\Delta = \{0\}$. ∎

Proposition 3.6 *If ϕ is ergodic, and $\hat{G}$ is torsion-free, then $\Delta = \Gamma$. In particular, ϕ has quasiperiodic spectrum iff it has quasidiscrete spectrum.*

Proof According to Theorem 3.3(iv), we have $\Delta_1 = \Gamma_1$. Suppose that $\Delta_n = \Gamma_n$, and let $\gamma \in \Delta_{n+1}$. Thus $\tau^{*k}\gamma - \gamma \in \Gamma_n$ for some $k > 0$. Let $\xi = \tau^*\gamma - \gamma$. Then

$$\tau^{*k}\xi - \xi = \tau^*(\tau^{*k}\gamma - \gamma) - (\tau^{*k}\gamma - \gamma) \in \Gamma_{n-1} \subseteq \Delta_{n-1}.$$

It follows that $\xi \in \Delta_n = \Gamma_n$, and hence that $\gamma \in \Gamma_{n+1}$. By induction $\Delta_n = \Gamma_n$ for all n, and $\Delta = \Gamma$. ▌

The following theorem shows that there was no loss in generality by assuming that τ was an automorphism in the preceding section.

Theorem 3.6 (*Seethoff*) *If ϕ has quasiperiodic spectrum, then it is invertible and distal.*

Proof By assumption $\Delta = \hat{G}$. Suppose that $\gamma \in \Delta_n$ for some $n > 0$. Then $\tau^{*k}\gamma - \gamma \in \Delta_{n-1}$. By iteration it follows that there is a polynomial p with constant term ± 1 such that $p(\tau^*)\gamma = 0$. Hence there is another polynomial q such that $\gamma = \tau^*[q(\tau^*)\gamma] = \tau^*\xi$. Thus τ^* is epic and hence invertible.

Following Seethoff [56], we divide the remainder of the proof into three parts.

I. *If $\Delta_1 = \hat{G}$, then τ is distal. Let*

$$\Pi_n = \{\gamma : \tau^{*n!}\gamma = \gamma\}.$$

Then each Π_n is a subgroup of $\hat{G}$, $\tau^*(\Pi_n) \subseteq \Pi_n$, $\Pi_n \subseteq \Pi_{n+1}$, and $\hat{G} = \Delta_1 = \bigcup_{n=1}^{\infty} \Pi_n$. Moreover, $\tau^{*n!}$ reduces to the identity on Π_n. It follows that the factor automorphism $\tau_n = \tau/\Pi_n^{\perp}$ on $G/\Pi_n^{\perp}$ (the adjoint of τ^* restricted to Π_n) also satisfies $\tau_n^{n!}$ = identity. But then it follows that τ_n is distal [$O_{\tau_n^2}(x, y)$ is a finite set]. Since (G, τ) is the inverse limit of the sequence $(G/\Pi_n^{\perp}, \tau_n)$, it is distal.

II. *If $\Delta_n = \hat{G}$, then τ is distal.* The proof is by induction. Suppose the statement is true for fixed n and all (G, τ). Suppose further that τ is an automorphism of G for which $\Delta_{n+1} = \hat{G}$. Let $H = \Delta_n^{\perp}$.

By the induction hypothesis the automorphism τ_1 induced by τ on G/H is distal, since $(G/H)\hat{} \cong \Delta_n(\tau)$ and $\tau_1{}^*$ is the restriction of τ^* to Δ_n. Moreover, the restriction τ_2 of τ to H is distal by I, since $\hat{H} \cong \hat{G}/\Delta_n = \Delta_{n+1}/\Delta_n = \Delta_1(\tau_2)$.

Now suppose that there is a net n_k of integers and points $x, y, z \in G$ with $\tau^{n_k}x \to z$, $\tau^{n_k}y \to z$. Let $w = x - y$. Then $\tau^{n_k}w \to 0$. It follows that $\tau_1^{n_k}(w + H) \to 0$. By distality of τ_1, $w + H$ is the zero coset, that is, $w \in H$. Since $\tau_2^{n_k}w = \tau^{n_k}w \to 0$, and τ_2 is distal, $w = 0$. Thus τ is distal.

III. *If $\Delta = \hat{G}$, then ϕ is distal.* Since Δ is the union of the increasing sequence Δ_n, it follows, as in the proof of I, that τ is the inverse limit of the distal transformations τ_n induced by τ on $G/\Delta_n{}^{\perp}$, hence distal.

Finally, suppose that $\phi^{n_k}(x) \to z$ and $\phi^{n_k}(y) \to z$. For any n we have

$$\phi^n(x) = \tau^n(x) + \tau^{n-1}(a) + \cdots + \tau(a) + a.$$

Thus $\phi^{n_k}(x) - \phi^{n_k}(y) = \tau^{n_k}(x) - \tau^{n_k}(y) = \tau^{n_k}(x - y) \to 0$. Hence $x - y = 0$. Thus ϕ is distal. ∎

We shall show in Chapter IV that the converse of Theorem 3.6 is also true. That is, ϕ is distal iff $\Delta = \hat{G}$. Thus it makes sense to talk about the distal part of τ, or at least the distal part of τ^*. The following theorem gives a precise meaning to the *ergodic part* of τ.

Theorem 3.7 (*Seethoff*) *Let τ be a continuous epimorphism of the compact abelian group G. Then there exists a uniquely determined closed subgroup H of G such that*

(i) $\tau(H) \subseteq H$,
(ii) *τ is ergodic on H, and*
(iii) *H contains every subgroup on which τ is ergodic.*

Moreover, τ/H has quasiperiodic spectrum.

Proof Set $H = \Delta^{\perp}$. For any closed τ-invariant subgroup F of G, let τ_F be the restriction of τ to F. The dual $\hat{F}$ of F is isomorphic to $\hat{G}/F^{\perp}$, and it is easily seen that

$$\Delta_n(\tau_F) = \Delta_n/F^{\perp}.$$

In particular, $\Delta_1(\tau_H) = \Delta_1/\Delta$ is trivial. According to Proposition 3.5, τ_H is ergodic. Conversely, if τ_F is ergodic, then $\Delta(\tau_F) = \Delta/F^{\perp}$ is trivial, so that $\Delta \subseteq F^{\perp}$. But this implies that $F \subseteq \Delta^{\perp} = H$.

Finally, $(G/H)\hat{} \cong H^{\perp} = \Delta$, so that τ/H has quasiperiodic spectrum. ∎

Combining Theorem 3.7 with Proposition 3.6 yields the following corollary.

Corollary 3.7.1 *Let ϕ be an ergodic affine transformation, $\phi(x) = \tau(x) + a$, on G. Suppose $\hat{G}$ is torsion-free. Then there exists a uniquely determined, closed, τ-invariant subgroup H of G such that*

(i) *τ restricted to H is ergodic, and*
(ii) *the transformation ϕ/H induced by ϕ on G/H has quasidiscrete spectrum.*

5. ERGODIC AUTOMORPHISMS

In the previous section we pointed out that an automorphism τ of a compact abelian group G has a distal part (quasiperiodic spectrum) and an ergodic part. Corollary 3.7.1 says the same for an ergodic affine transformation on a connected group, with the distal part having quasidiscrete spectrum. Theorem 3.5 gives a concrete description of ergodic affine transformations with quasidiscrete spectrum. In this section we shall prove an analogous theorem for ergodic automorphisms and look at some examples.

In order to introduce the class of ergodic automorphisms τ for which our representation theorem is valid, we first prove the following proposition.

Proposition 3.7 *Let G be a compact abelian metric group, and let τ be a continuous, ergodic automorphism of G. Then there exists an $a \in G$ such that the orbit $O_\tau(a)$ of a under τ separates the points of $\hat{G}$, that is, $\gamma(\tau^n a) = 1$ $(n = 0, \pm 1, \pm 2, \ldots)$ implies that $\gamma = 0$.*

Proof We shall in fact show that for "most" of the points $a \in G$ the orbit of a is dense in G.

Since G is compact, it is totally bounded. Thus there exists a double sequence $\{B_{nk}\}$ of open balls in G such that $G \subseteq \bigcup_{n,k} B_{nk}$ and the radius of B_{nk} is $1/n$ for $k = 1, \ldots, k_n$ and $n = 1, 2, \ldots$. Let G_{nk} denote the set of $x \in G$ for which $O_\tau(x) \cap B_{nk} = \varnothing$. The set $G_0 = \bigcup_{n,k} G_{nk}$ is exactly the set of points whose orbits are *not* dense in G. Since we can write G_{nk} as

$$G_{nk} = \bigcap_{n=-\infty}^{\infty} \{x \in G : \tau^n(x) \notin B_{nk}\},$$

and since τ is continuous, each G_{nk} is closed. Moreover, if we let A_{nk} denote the orbit of G_{nk},

$$A_{nk} = \{\tau^j(x) : x \in G_{nk},\, n \in Z\},$$

then A_{nk} is τ-invariant and disjoint from B_{nk}.

Since B_{nk} is open and hence has positive Haar measure, it follows from the ergodicity of τ that A_{nk} and hence $G_{nk} \subseteq A_{nk}$ have measure zero. Therefore G_{nk} has empty interior.

We have shown that G_0 is a countable union of closed, nowhere dense sets. By the Baire category theorem, G_0 cannot be all of G. In fact, the set of points with dense orbits is a dense G_δ. ∎

For an arbitrary continuous automorphism τ, let us introduce the following notation. By $\mathscr{E}(\tau)$ we shall mean the semigroup of endomorphisms ϕ of G of the form

$$\phi(x) = \sum_{j=1}^{l} n_j \tau^{k_j}(x),$$

where $n_1, \ldots, n_l, k_1, \ldots, k_l \in Z$ and $l \in Z^+$. By Gp Orb(a) or Gp Orb$_\tau(a)$ we shall mean the group generated by the orbit $O_\tau(a)$. Clearly, Gp Orb$_\tau(a) = \mathscr{E}(\tau)a$.

Definition 3.6 The affine system (G, τ) is *monothetic* if there exists an $a \in G$ such that Gp Orb$_\tau(a) = \mathscr{E}(\tau)a$ is dense in G.

Remark If we denote the identity automorphism on G by I, then (G, I) is monothetic iff G is a monothetic group in the usual sense, that is, iff a is a topological generator of G.

Proposition 3.8 *The system (G, τ) is monothetic iff there exists an $a \in G$ such that $O_\tau(a)$ separates the points of $\hat{G}$.*

The proof is left to the exercises.

We are now in a position to offer the analog of Theorem 3.5 for ergodic automorphisms. Let K_{d}^{ω} be the discretized version of the infinite-dimensional torus $K^{\omega} = \bigotimes_{n=-\infty}^{\infty} K$. That is, K_{d}^{ω} is K^{ω} with the discrete topology.

Note that K_{d}^{∞} is a proper subgroup of K_{d}^{ω}, since K_{d}^{∞} contains only the finitely nonzero sequences from K, whereas K_{d}^{ω} contains all bisequences. Let $\tilde{\tau}^*$ be the shift transformation on K_{d}^{ω}, that is, $\tilde{\tau}^*(z) = w$, where $w_n = z_{n+1}$, Then $\tilde{\tau}^*$ is an automorphism of K_{d}^{ω}, and its adjoint $\tilde{\tau}$ is a continuous automorphism of the dual group $\overline{Z^{\infty}} = (K_{\mathrm{d}}^{\omega})\hat{}$. According to Theorem 3.7, there is a maximal τ-invariant subgroup H of $\overline{Z^{\infty}}$ such that the restriction $\tilde{\tau}_H$ of $\tilde{\tau}$ to H is ergodic. Let Φ_{e} denote the ergodic system $(H, \tilde{\tau}_H)$.

Remark $\overline{Z^{\infty}}$ is the Bohr compactification of the discrete group Z^{∞} (see [53]).

Theorem 3.8 *If τ is an ergodic automorphism of the compact abelian group G, and if (G, τ) is monothetic, then (G, τ) is an algebraic factor of Φ_{e}.*

Proof As in the proof of Theorem 3.5, we define a mapping ρ^* of $\hat{G}$ into $\hat{H} = K_{\mathrm{d}}^{\omega}/\Delta$, where $\Delta = \Delta(\tilde{\tau}) = H^{\perp}$ is defined as in Section 4, and show that (i) ρ^* is monic and (ii) $\rho^*\tau^* = \tilde{\tau}_H{}^*\rho^*$. The adjoint $\rho: H \to G$ is then an algebraic homomorphism of $(H, \tilde{\tau}_H)$ onto (G, τ).

By assumption there exists $a \in G$ such that $O_\tau(a)$ separates the points of $\hat{G}$. For $\gamma \in \hat{G}$ define $\rho_0{}^*(\gamma)$ to be the bisequence

$$\rho_0{}^*(\gamma) = \{\gamma(\tau^n a)\}. \tag{18}$$

Then $\rho_0{}^*\colon G \to K_{\mathrm{d}}{}^\omega$ is clearly monic and satisfies $\rho_0{}^*\tau^* = \tilde{\tau}^*\rho_0{}^*$. Moreover,

$$\rho_0{}^*(\hat{G}) \cap \Delta(\tilde{\tau}) = \{0\}. \tag{19}$$

For suppose that $z = \rho_0{}^*(\gamma) \in \Delta_p$ for some $p > 0$. Then there exists a positive integer k with $\tilde{\tau}^{*k}z - z \in \Delta_{p-1}$. This means that

$$\begin{aligned}\tilde{\tau}^{*k}z - z &= \{\gamma(\tau^{n+k}a)\} - \{\gamma(\tau^n a)\} = \{\gamma(\tau^{n+k}a)/\gamma(\tau^n a)\}\\ &= \{\tau^{*k}\gamma(\tau^n a)/\gamma(\tau^n a)\} = \rho_0{}^*(\tau^{*k}\gamma - \gamma) \in \Delta_{p-1}.\end{aligned} \tag{20}$$

Suppose that $\rho_0{}^*(\hat{G}) \cap \Delta_{p-1} = \{0\}$. Then (20) implies that $\tau^{*k}\gamma - \gamma = 0$. Since τ is ergodic, this implies by Theorem 3.1 that $\gamma = 0$. We conclude that

$$\rho_0{}^*(\hat{G}) \cap \Delta_{p-1} = \{0\} \Rightarrow \rho_0{}^*(\hat{G}) \cap \Delta_p = \{0\}.$$

Since $\rho_0{}^*$ is monic, $\rho_0{}^*(\hat{G}) \cap \Delta_0 = \{0\}$, and (19) follows by induction on p and the fact that $\Delta = \bigcup_{p=0}^{\infty} \Delta_p$.

Finally, let $\rho^*\colon \hat{G} \to K_{\mathrm{d}}{}^\omega/\Delta$ be the composition of $\rho_0{}^*$ and the natural projection π of $K_{\mathrm{d}}{}^\omega$ onto $K_{\mathrm{d}}{}^\omega/\Delta$. It follows then from (19) that ρ^* is monic. Moreover, $\pi\tilde{\tau}^* = \tilde{\tau}_H{}^*\pi$ by the definition of factor automorphism. Thus

$$\rho^*\tau^* = \pi\rho_0{}^*\tau^* = \pi\tilde{\tau}^*\rho_0{}^* = \tilde{\tau}_H{}^*\pi\rho_0{}^* = \tilde{\tau}_H{}^*\rho^*,$$

and we are finished. ▌

Corollary 3.8.1 *Let $B = \rho_0{}^*(\hat{G})$. Then B is a subgroup of $K_{\mathrm{d}}{}^\omega$ with the following properties:*

1. $\tilde{\tau}^*(B) = B$, *and*
2. $B \cap \Delta(\tilde{\tau}) = \{0\}$.

Conversely, corresponding to any subgroup B of $K_{\mathrm{d}}{}^\omega$ satisfying properties 1 and 2 there is a compact abelian group G and a continuous, ergodic automorphism τ of G such that (G, τ) is monothetic and $B = \rho_0{}^*(\hat{G})$, where $\rho_0{}^*$ is defined by (18).

Proof The first statement has already been proved. To prove the converse, let $G = \hat{B} = \overline{Z^\infty}/B^\perp$. From property 1 it follows that $\tilde{\tau}(B^\perp) = B^\perp$ and that the adjoint of the restriction of $\tilde{\tau}^*$ to B is the factor automorphism $\tilde{\tau}/B^\perp$ on $\overline{Z^\infty}/B^\perp$. Let us denote $\tilde{\tau}/B^\perp$ by τ.

Let $a_0 \in \overline{Z^\infty} = (K_{\mathrm{d}}{}^\omega)\hat{}$ be defined by $\langle a_0, z\rangle = z_0$ for $z \in K_{\mathrm{d}}{}^\omega$, and let a be the projection of a_0 on $G = \overline{Z^\infty}/B^\perp$. Since $\langle \tilde{\tau}^n a_0, z\rangle = \langle a_0, \tilde{\tau}^{*n}z\rangle = z_n$, it follows that $O_{\tilde{\tau}}(a_0)$ separates the points of K_d^{ω}, and hence $O_{\tau}(a)$ separates the points of $B = \hat{G}$. Thus (G, τ) is monothetic.

If ρ_0^* is defined by (18), then $\rho_0^*(z) = z$ for each $z \in B = \hat{G}$. It only remains to show that τ is ergodic. But this follows immediately from property 2 and Theorem 3.1. ▮

An interesting side issue is the question of which compact abelian groups admit monothetic automorphisms. Let us agree to call such a group *algebraically monothetic*. Thus G is algebraically monothetic iff there exist $a \in G$ and a continuous automorphism τ of G such that $\mathscr{E}(\tau)a$ is dense in G. Corollary 3.8.1 provides us with one answer to the above question.

Corollary 3.8.2 *The group G is algebraically monothetic iff $\hat{G}$ is isomorphic to a (not necessarily closed) subgroup B of the infinite-dimensional torus K^{ω} satisfying $\tilde{\tau}^*(B) = B$, where $\tilde{\tau}^*$ is the shift on K^{ω}.*

Proof The condition $B \cap \Delta(\tilde{\tau}) = \{0\}$ was needed only to prove ergodicity. Otherwise, the proof proceeds as above. ▮

Example 7 For any $\alpha \in K_d$ denote by (α) the cyclic group generated by α, and let $B = \bigotimes_{n=-\infty}^{\infty} (\alpha)$ be the complete direct product. B is identified with a subgroup of K_d^{ω} in the obvious way and satisfies property 1. However, property 2 is not satisfied, since, for example, $z = \{\alpha^{k_n}\} \in \Delta$ if the integers k_n satisfy the recurrence relation $k_{n+j} = k_n$ $(n = 0, \pm 1, \pm 2, \ldots)$.

On the other hand, the subgroup $B^1 = \bigoplus_{n=-\infty}^{\infty} (\alpha)$ of B, consisting of sequences $z = \{\alpha^{k_n}\}$ with $k_n = 0$ for $|n| > N(z)$, satisfies both of properties 1 and 2. (Recall that $\Delta_1(\tau) = \{0\}$ implies $\Delta(\tau) = \{0\}$.)

Example 8 More generally, let $B = \bigoplus_{n=-\infty}^{\infty} T$, where T is any subgroup of K_d. Then conditions 1 and 2 are satisfied. In this case, $\hat{B} = \bigotimes_{n=-\infty}^{\infty} \hat{T}$, and τ is the *symbolic flow* on $\hat{T}$, as defined in Example 3 of Section 3.1. If we take T to be the kth roots of unity, τ is the Bernoulli shift on k points.

Corollary 3.8.3 *The symbolic flow on any monothetic group is monothetic. In particular, Bernoulli shifts are monothetic.*

Example 9 The divisibility of the group K_d and hence of K_d^{ω} implies by an easy induction argument that Δ is divisible. It follows from a well-known theorem of group theory [38, p. 8] that Δ is a direct summand of K_d^{ω}. That is, there exists a group B_e such that

$$K_d^{\omega} = \Delta \oplus B_e.$$

Clearly, B_e satisfies condition 2. We would like to say that B_e can be chosen to also satisfy condition 1, for this would imply that the "model" Φ_e of Theorem 3.8 is monothetic since $B_e \cong \hat{H}$. It would also imply that any system (G, τ), where τ is monothetic, has a direct sum decomposition $G = G_1 \oplus G_2$ (since the model $\overline{Z^\infty}$ would decompose as $H \oplus \hat{B}_e$) with G_1 and G_2 τ-invariant, (G_1, τ) ergodic, and (G_2, τ) distal. However, this is *not true*, as shown by a counterexample constructed by Kerrick [39] on the 3-torus K^3.

Remark In Examples 7 and 8, the group $B = \rho_0{}^*(G)$ is a subgroup of the direct sum $K_d{}^\infty \subseteq K_d{}^\omega$. This, however, does not exhaust the possibilities. It can be shown (Exercises 29 and 30) that any automorphism of K^n is monothetic. Kerrick has shown [39, p. 55] that (in his Type I case) a can be chosen so that $B = \rho_0{}^*(Z^n)$ is given by

$$B = \left\{\left\{\exp 2\pi i \left[\sum_{j=1}^{r} \alpha_j \operatorname{Re}(\lambda_j{}^n(m \cdot u_j))\right]\right\} : m \in Z^n\right\}, \tag{21}$$

where the λ_j are eigenvalues and the u_j eigenvectors of the associated unimodular matrix, and where the α_j are real numbers. It can be shown [13] that not all such B are contained in $K_d{}^\infty$. Another example is constructed in [13] of a nontrivial B satisfying conditions 1 and 2 and having trivial intersection with $\Delta \oplus K_d{}^\infty$.

6. AN AFFINE TRANSFORMATION ASSOCIATED WITH THE DYNAMICAL SYSTEM Φ

The time has arrived to pull together ideas from the last three chapters and discuss the affine system associated with an abstract dynamical system $\Phi = (X, \mathscr{B}, \mu, \phi)$ or with a classical system $\Phi = (X, \phi)$.

Let us denote by $\mathscr{G}$ or $\mathscr{G}(\Phi)$ the group of all complex-valued functions on X which have constant absolute value one and which, in the abstract case, are $\mathscr{B}$-measurable or, in the classical case, are continuous. In the abstract case, we shall actually deal with equivalence classes of such functions, two functions being equivalent if they are equal μ-almost everywhere. In this case, $\mathscr{G} \subseteq L_2(X)$. In the classical case, $\mathscr{G} \subseteq C(X)$. It is clear that $\mathscr{G}$ is a group, and that the set K_1 of constant functions of modulus one is a subgroup.

As explained in Chapters I and II, the dynamical system Φ has associated with it a bounded linear operator T_ϕ on $L_2(X)$ or on $C(X)$, defined by $T_\phi f(x) = f(\phi(x))$.

Moreover, the restriction of this operator to $\mathscr{G}$ is a monomorphism of the group $\mathscr{G}$. If Φ is invertible, it is an automorphism. We shall denote this transformation of $\mathscr{G}$ by τ^* or $\tau_\Phi{}^*$. Thus

$$\tau_\Phi{}^* f(x) = T_\phi f(x) = f(\phi(x)). \tag{22}$$

The mapping $\alpha_0 : K_1 \to K$, which assigns to each constant function its constant value, is clearly a homomorphism. Since K is divisible, α_0 has a homomorphic extension $\alpha: \mathscr{G} \to K$.

Let us define an endomorphism $\sigma^*: \mathscr{G} \to \mathscr{G}$ by $\sigma^* f = \tau^* f/f$. This is in keeping with our earlier notation, bearing in mind that the group $\mathscr{G}$ is perforce written multiplicatively. Let us denote $\hat{G} = \sigma^*(\mathscr{G})$.

Note The mappings τ^*, σ^*, α here are the same as $\tilde{\tau}$, $\tilde{\sigma}$, α introduced in the proof of Proposition 3.4 and Theorem 3.4 of Section 3.

Proposition 3.9 *If Φ is ergodic (or minimal), then $\mathscr{G} \cong K \oplus \hat{G}$.*

Proof The mapping $\psi: f \to (\alpha(f), \sigma^*(f))$ is a homomorphism of $\mathscr{G}$ into $K \oplus \hat{G}$. It is epic because the range of σ^* is $\hat{G}$, and, for a fixed $f \in \mathscr{G}$ and $\lambda \in K$, it maps $g = [\lambda/\alpha(f)]f$ into $(\lambda, \sigma^*(f))$. It is monic because

$$\sigma^* f = 1 \Rightarrow \tau^* f = T_\phi f = f \Rightarrow f \in K_1,$$

and, for $f \in K_1$ with $\alpha(f) = 1$, we have $f = 1$. ∎

Remark The inverse map ψ^{-1} carries K onto K_1 and $\hat{G}$ onto the kernel of α.

We shall assume from now on that Φ is ergodic or minimal, so that Proposition 3.9 applies. The notation $\hat{G} = \sigma^*(\mathscr{G})$ was chosen with a purpose in mind. The compact dual G of the group $\hat{G}$ is precisely the group on which the affine transformation $\tilde{\phi}$ associated with Φ will be defined.

By construction $\alpha \in \hat{\mathscr{G}}$ is a character on $\mathscr{G}$. Thus it determines a character on the subgroup $\hat{G} = \sigma^*(\mathscr{G})$. We shall denote this again by α. Thus $\alpha \in G$. Since τ^* commutes with σ^*, it is clear that $\tau^*\hat{G} \subseteq \hat{G}$, with equality holding if Φ is invertible. Thus τ^* has an adjoint τ on G, which is an epimorphism in general and an automorphism if Φ is invertible. Finally, we define the affine transformation $\tilde{\phi}$ of G by

$$\tilde{\phi}(\xi) = \tau(\xi) + \alpha. \tag{23}$$

The following result due to Abramov is basic to the representation theory to follow. The subgroups A and B of $\mathscr{G}$ are defined as in Section 3. Thus $A_0 = \{1\}$, $B_0 = K_1$, $B_{n+1} = \{f \in \mathscr{G} : \sigma^* f \in B_n\}$, $A_{n+1} = \sigma^*(B_{n+1})$ for $n = 0, 1, 2, \ldots$, and $A = \bigcup_{n=0}^\infty A_n$, $B = \bigcup_{n=0}^\infty B_n$.

Lemma 3.1 *Suppose $f, g \in B$ with $\sigma^*f \neq \sigma^*g$. Further suppose that T_ϕ has no eigenvalues that are roots of unity. Then $(f, g) = 0$; that is, f and g are orthogonal in $L_2(X)$. In the classical case, the integral may be calculated with respect to any ϕ-invariant, ergodic Borel measure on X.*

Proof The isomorphism $\psi: f \to (\alpha(f), \sigma^*(f))$ of Proposition 3.9 carries B_n onto $K \oplus A_n$ for each n and B onto $K \oplus A$. It follows from the remark following that proposition that

$$B_n = K_1 \oplus C_n,$$

where $C_n = B_n \cap \alpha^{-1}(1)$. Let us denote by $\mathscr{X}_n$ the closed linear subspace of $L_2(X, \mathscr{B}, \mu)$ generated by C_n. Following Abramov [2], we shall show first that $C_{n+1} \sim \mathscr{X}_n \subseteq \mathscr{Y}_n$, where $\mathscr{Y}_n = \mathscr{X}_n^\perp$ is the orthogonal complement of $\mathscr{X}_n$ in the Hilbert space $L_2(X, \mathscr{B}, \mu)$.

Let $f \in C_{n+1} \sim \mathscr{X}_n$ have the unique representation $f = f_1 + f_2, f_1 \in \mathscr{X}_n$, $f_2 \in \mathscr{Y}_n$. Since $T_\phi(B_n) \subseteq B_n \subseteq \mathscr{X}_n$, it follows that $T_\phi(\mathscr{X}_n) \subseteq \mathscr{X}_n$. Thus $T_\phi f_1 \in \mathscr{X}_n$. Moreover, since T_ϕ is invertible on B_n (cf. the proof of Theorem 3.6), it follows that T_ϕ is unitary on $\mathscr{X}_n$. Thus for each $g \in \mathscr{X}_n$,

$$(T_\phi f_2, g) = (f_2, T_\phi^{-1} g) = 0.$$

That is, $T_\phi f_2 \in \mathscr{Y}_n$.

Since $f \in C_{n+1}$, we have

$$T_\phi f = \psi^* f = \sigma^* f \cdot f = \sigma^* f \cdot f_1 + \sigma^* f \cdot f_2, \tag{24}$$

as well as

$$T_\phi f = T_\phi f_1 + T_\phi f_2. \tag{25}$$

Since $\sigma^* f \in A_{n+1} \subseteq B_n$, we have $\sigma^* f \cdot f_1 \in \mathscr{X}_n$, and as before

$$(\sigma^* f \cdot f_2, g) = (f_2, \overline{\sigma^* f} \cdot g) = 0$$

for each $g \in \mathscr{X}_n$, so that $\sigma^* f \cdot f_2 \in \mathscr{Y}_n$. From (24) and (25) and the uniqueness of the orthogonal decomposition of the function $T_\phi f$, we conclude that

$$T_\phi f_1 = \sigma^* f \cdot f_1, \qquad T_\phi f_2 = \sigma^* f \cdot f_2. \tag{26}$$

From (26) and the definition of B_{n+1}, we conclude that either $f_j = 0$ a.e. or $g_j = f_j / |f_j| \in B_{n+1}$ with $\sigma^*(g_j) = \sigma^*(f)$ $(j = 1, 2)$. In the latter case $\sigma^*(f/g_j) = 1$, so that $f = \lambda_j g_j$, $\lambda_j \in K_1$. Since $f \notin \mathscr{X}_n$, it follows that $f_1 = 0$, and $f = f_2 \in \mathscr{Y}_n$, as asserted.

Next let us show that $(f, g) = 0$ for all $f, g \in C_1$, $f \neq g$. This follows from

$$(f, g) = (T_\phi f, T_\phi g) = (\sigma^* f)(\sigma^* g)^{-1}(f, g), \tag{27}$$

where $\sigma^* f$, $\sigma^* g \in K_1$, are constants. Since $f \neq g$ and σ^* is monic on C_n, it follows from (27) that $(f, g) = 0$.

Suppose now that $(f, g) = 0$ for all $f, g \in C_n$, $f \neq g$. We refine our earlier result by showing that $C_{n+1} \sim C_n \subseteq \mathscr{Y}_n$. Equivalently, we show that $C_{n+1} \cap \mathscr{X}_n \subseteq C_n$. Suppose then that $f \in C_{n+1} \cap \mathscr{X}_n$. By the induction hypothesis, the elements of C_n form an orthonormal basis for $\mathscr{X}_n$. Thus

$$\begin{aligned} f &= \sum_{g \in C_n} (f, g)g, \\ \|f\|_2^2 &= \sum_{g \in C_n} |(f, g)|^2 < \infty. \end{aligned} \tag{28}$$

For each $g \in C_n$ we have $T_\phi g = \sigma^*(g)g$, and since $\sigma^*(g) \in B_{n-1} = K \oplus C_{n-1}$, $T_\phi g = cg_1$, where $|c| = 1$ and $g_1 g^{-1} \in C_{n-1}$. Likewise, $\sigma^*(f) = ah$, where $|a| = 1$ and $h \in C_n$. Thus

$$\begin{aligned} (f, g) &= (T_\phi f, T_\phi g) = (ahf, cg_1) \\ |(f, g)| &= |(f, h^{-1} g_1)|. \end{aligned}$$

Repeating this argument, we get an infinite sequence $g_n \in C_n$ with $g_n g^{-1} \in C_{n-1}$ and

$$|(f, g)| = |(f, h^{-n} g_n)|.$$

From (28) it follows that only a finite member of the functions $h^{-n} g_n \in C_n$ are distinct. Thus for some n, $h^{-n} g_n = g$, and so $h^n = g_n g^{-1} \in C_{n-1}$. Now since $C_n / C_{n-1} \cong A_n / A_{n-1} \cong \sigma^{*n-1}(A_n) \subseteq \sigma^*(B_1) = A_1$ (see the proof of Proposition 3.4), and since A_1 consists of the eigenvalues of T_ϕ and thus has no elements of finite order, it follows that $h \in C_{n-1}$. But this means that $\sigma^*(f) = ah \in B_{n-1}$. Therefore $f \in B_n \cap C_{n+1} = C_n$, as asserted.

Now let $f, g \in C_{n+1}$ with $f \neq g$. If both $f, g \in C_n$, then $(f, g) = 0$ by the induction hypothesis. If $f \notin C_n$ and $g \in C_n$, then $(f, g) = 0$ by what was just proved. Thus the induction step is reduced to considering the case $f, g \in C_{n+1} \sim C_n$. In this case,

$$(f, g) = (T_\phi f, T_\phi g) = (\sigma^*(f) f, \sigma^*(g) g) = (f\bar{g}, \overline{\sigma^*(f)}\, \sigma^*(g)).$$

Since $\overline{\sigma^*(f)}\, \sigma^*(g) \in B_n$, there exist a constant c with $|c| = 1$ and $h \in C_n$ with $\overline{\sigma^*(f)}\, \sigma^*(g) = ch$. Since $f\bar{g} \in C_{n+1}$, either

$$(f, g) = (f\bar{g}, ch) = 0,$$

or $f\bar{g} \in C_n$ and $f\bar{g} = h$. But the latter implies that

$$cf\bar{g} = \overline{\sigma^*(f)}\, \sigma^*(g)$$
$$cT_\phi f = c\sigma^*(f)f = \sigma^*(g)g = T_\phi g$$
$$f = g,$$

a contradiction. Thus $(f, g) = 0$.

Finally, suppose that $f, g \in B$ with $\sigma^*(f) \neq \sigma^*(g)$. Then $f = cf_1$, $g = dg_1$ with $|c| = |d| = 1$ and $f_1, g_1 \in C = \bigcup_{n=1}^{\infty} C_n$. Since $\sigma^*(f) = \sigma^*(f_1)$, $\sigma^*(g) = \sigma^*(g_1)$, it follows that $f_1 \neq g_1$. Thus $(f, g) = cd^{-1}(f_1, g_1) = 0$. ∎

Remark If Φ is a totally ergodic (totally minimal) abstract (classical) dynamical system, then T_ϕ can have no eigenvalues that are roots of unity. (Exercise 35.)

Theorem 3.9 *Suppose that Φ is totally ergodic or totally minimal. Then $\tilde{\phi}_\Phi = \tilde{\phi}$, defined by Eq. (23), is ergodic. Moreover,*

(i) *τ_Φ is ergodic iff ϕ has no nontrivial quasieigenfunctions, and*

(ii) *(Abramov–Hahn–Parry) if ϕ has quasidiscrete spectrum, then $\tilde{\phi}$ also has quasidiscrete spectrum, and they are isomorphic in the appropriate category.*

Proof (*Ergodicity*) According to Theorem 3.3(iv), it will suffice to show that

$$\tau^{*n}f = T_\phi^n f = f \Rightarrow \tau^* f = T_\phi f = f$$

and

$$T_\phi f = f, \qquad \alpha(f) = 1 \Rightarrow f = 1$$

for each $f \in \hat{G} = \sigma^*(\mathcal{G})$. These implications are in fact valid for all $f \in \mathcal{G}$. The first implication follows from the ergodicity or minimality of ϕ^n, since $T_\phi f = f$ for those (and only those) $f \in K_1$. The second implication follows from the definition of α, since $\alpha(f) = f$ for $f \in K_1$.

(i) Using the results from Section 4, we see that the ergodic part of $\tilde{\phi}$ is defined on $H = \Delta^\perp = \Gamma^\perp$ where Γ is the set of functions $g \in \hat{G}$ such that $\sigma^{*n} g = 1$ for some n. Thus H is all of G iff Γ is trivial, that is, iff $\sigma^{*(n+1)}f = \sigma^{*n}(\sigma^* f) = \sigma^{*n} g = 1$ for $f \in \mathcal{G}$ implies $g = 1$. By ergodicity this is equivalent to $f \in K_1$.

(ii) Suppose that Φ has quasidiscrete spectrum. In the abstract case, this means that the subgroup B of $\mathcal{G}$ consisting of all quasieigenfunctions of ϕ spans $L_2(X)$ (Definition 3.3). This implies in turn that the only continuous linear functional on $L_2(X)$ that annihilates B is the zero functional. Let $\xi \in H = \Delta^{\perp}(\tau) = A^{\perp} \subseteq G$. Define $\lambda: \mathscr{G} \to C$ by

$$\lambda(f) = \alpha(f)\,[\xi(\sigma^* f) - 1]. \tag{29}$$

We shall show that λ vanishes on B and that λ can be extended to a linear functional on $L_2(X)$.

Let L be the linear span in $L_2(X)$ of B. For each $f \in B$ there is exactly one $f_1 \in B$ such that $\sigma^* f = \sigma^* f_1$ and $\alpha(f_1) = 1$, namely, $f_1 = \alpha(f)^{-1} f$. Moreover, $\lambda(f_1) = \alpha(f)^{-1}\,\lambda(f)$. Thus each $f \in L$ has a representation

$$f = c_1 f_1 + c_2 f_2 + \cdots + c_n f_n, \tag{30}$$

where the c_j are complex numbers and $\sigma^* f_i \neq \sigma^* f_j$ for $i \neq j$. According to Lemma 3.1, this representation is unique, and the formula

$$\lambda(f) = c_1 \lambda(f_1) + c_2 \lambda(f_2) + \cdots + c_n \lambda(f_n) \tag{31}$$

determines a bounded linear functional on L. By the Hahn–Banach theorem, λ has an extension $\tilde{\lambda}$ to a continuous linear functional on $L_2(X)$.

For $f \in B$, $\sigma^*(f) \in A$, so that $\xi(\sigma^* f) = 1$, and

$$\tilde{\lambda}(f) = \lambda(f) = \alpha(f)\,[1 - 1] = 0.$$

It follows that $\lambda = 0$, and from (29) that $\xi(g) = 1$ for all $g \in \hat{G} = \sigma^*(\mathscr{G})$. Thus H is trivial, $\Delta(\tau) = \hat{G}$, and $\tilde{\phi}$ has quasidiscrete spectrum.

To complete the proof, we need to show that Φ is measure-theoretically isomorphic to $\tilde{\Phi} = (G, \tilde{\phi})$, where $G = \hat{A}$. First, we use the lemma again to define a linear mapping S from the linear span L of B in $L_2(X)$ to $L_2(G)$. Namely,

$$S\left(\sum_{j=1}^{n} c_j f_j\right)(\xi) = \sum_{j=1}^{n} c_j \xi(\sigma^* f_j), \qquad \xi \in G, \tag{32}$$

where the f_j are chosen to belong to the kernel of α. That is, $f \in \ker(\alpha)$ is mapped into $\sigma^*(f)$, and S is extended by linearity. Since distinct elements of $\hat{G}$ are clearly orthogonal in $L_2(G)$, it follows that S is an invertible isometry of L onto a dense subset of $L_2(G)$. Let $\tilde{S}$ be the unique extension of S to $L_2(X)$.

Finally, we need to show that $\tilde{S} = T_\rho$ for some measure-preserving map ρ of G onto X. As in Proposition 1.2, this will be proved by showing that $\tilde{S}$ is doubly stochastic.

Since $\tilde{S}^* = \tilde{S}^{-1}$, it is clear from (32) that $\tilde{S}1 = 1$ and $\tilde{S}^*1 = 1$. Let us show that $Sf \geq 0$ for all $f \geq 0$.

Since S restricted to B is a group homomorphism into G, it follows that

$$S(fg) = (Sf)(Sg) \tag{33}$$

holds for all f, $g \in B$. By linearity (33) holds for all f, $g \in L$. Let $F \in \mathscr{B}$ and choose $g_n \in L$ with $g_n \to \chi_B$ in $L_2(X)$. For each $f \in L$, both f and Sf are bounded functions on X and G, respectively. Therefore, $fg_n \to f\chi_F$ in $L_2(X)$ and $(Sf)(Sg_n) \to (Sf)(S\chi_F)$ in $L_2(G)$. Thus $S(f\chi_F) = (Sf)(S\chi_F)$. Similarly, letting $f_n \to \chi_F$ gives $S(\chi_F) = S(\chi_F{}^2) = (S\chi_F)^2$. Therefore, $S\chi_F \geq 0$. It follows that $Sf \geq 0$ for all $f \geq 0$.

The proof of (ii) in the abstract case is now completed as in Exercise 1.6. ▌

Remark In the case of a totally minimal, classical dynamical system, the proof of (ii) is completed by noting that $S \geq 0$, $S1 = 1$ implies that S carries bounded functions into bounded functions and has norm one as an operator from L into $C(G)$. Moreover, a point transformation that carries all continuous functions on X into continuous functions on G must be continuous.

EXERCISES

Affine Transformations

1. Let $\tau: G \to G$ be a continuous epimorphism. Define a measure μ on G by

$$\mu(A) = m(\tau^{-1}(A)).$$

Show that μ is translation invariant and hence, by uniqueness of Haar measure, must be m. Show by example that τ is not necessarily measure preserving if it is only assumed to be an endomorphism (not epic).

2. If $T = (t_{ij})$ is an $n \times n$ matrix with integer entries, then $\tau^*(\gamma) = \gamma T$ defines an endomorphism of Z^n. Show that τ^* is the adjoint of the mapping τ of Example 2. If T has nonzero determinant, it determines a nonsingular transformation of R^n whose restriction to Z^n is τ^*. Deduce that τ^* is monic and hence τ is epic. Show that the τ corresponding to $T = \left(\begin{smallmatrix}1 & 1\\ 1 & 1\end{smallmatrix}\right)$ is not epic.

3. **(a)** Show that σ as defined in Example 3 is a bicontinuous group automorphism.

(b) Show that the formulas for σ^* and τ^* given in Examples 3 and 4 are correct.

4. Let G_0, G, σ, and τ be as in Examples 3 and 4. Let $a_0 \in G_0$ and define $a \in G$ to have for its nth component $\delta_{n0}\, a_0$. Let $H_0 \subseteq G$ be the subgroup

$$H_0 = \{x \in G : x_n = 0 \text{ for } n \leq 0\},$$

and $K_0 \subseteq G$ the affine subspace $K_0 = a + H_0$.

(a) Show that K_0 is invariant under σ [i.e., $\sigma(K_0) \subseteq K_0$] and hence under τ.

(b) Show that the restriction of τ to K_0 is *isomorphic* to the transformation ϕ of Example 5; that is, find an invertible affine mapping ψ between the appropriate spaces such that $\phi\psi = \psi\tau$.

5. If G is *divisible* as well as compact, then tx is defined in G for each $x \in X$ and each real number t. Suppose that G is divisible and ϕ is affine, that is, $\phi(x) = \tau(x) + a$ for some endomorphism τ of G. Show that ϕ is affine in the sense of convex sets, that is,

$$\phi(sx + ty) = s\phi(x) + t\phi(y) \qquad \text{for} \quad x, y \in G,\ s, t \geq 0,\ s + t = 1. \quad (34)$$

Conversely, show that if (34) holds for all x, $y \in G$ and $s = t = \frac{1}{2}$, then ϕ is affine in the former sense with $a = \phi(0)$.

6. Verify statements (i), (ii), and (iii) in Example 6.

Ergodicity

7. According to Theorem 3.2, a finite group admits no ergodic automorphisms. Can you give a simpler proof of this fact? Show that a group automorphism (of any compact group) is *never* minimal.

8. If T is a 3×3 unimodular matrix, then the characteristic equation of T is

$$P(\lambda) = \lambda^3 - s\lambda^2 + t\lambda \pm 1 = 0.$$

The roots are either all real or one real and a complex conjugate pair. Consider all cases and give a complete description of the ergodic case in terms of s, t and the determinant. [Hint: If τ is ergodic, then $P(\lambda)$ is not factorable over Z. Otherwise, there would be a linear factor with integer coefficients, making ± 1 an eigenvalue. If λ_0 is a root of unity satisfying $P(\lambda_0) = 0$, then the minimal polynomial of λ_0 divides (hence equals) $P(\lambda)$. What are the minimal polynomials of degree 3 of roots of unity?]

9. **(a)** If σ is defined as in Example 3, then $\sigma^{*k}(\gamma) = \gamma$ means $\gamma_{n+k} = \gamma_n$ for all n. Use the fact that only finitely many of the γ_n are nonzero to conclude that $\gamma = 0$ and σ is ergodic.

(b) Let τ be as in Example 4. Then $\tau^{*k}(\gamma) = \delta$, where $\delta_n = \sum_{j=0}^{k} \binom{k}{j} \gamma_{n+j}$. Assume that $\tau^{*k}(\gamma) = \tau^{*l}(\gamma)$ for some $0 \le k < l$ and $\gamma \ne 0$. Let p be the smallest integer with $\gamma_p \ne 0$ and set $n + l = p$ to reach a contradiction. Thus τ is ergodic.

10. Let ϕ_0 be a totally ergodic measure-preserving transformation of [0, 1]. Define ϕ on $[0, 1] \cup [2, 3]$ by

$$\begin{aligned} \phi(x) &= \phi_0(x) + 2 \qquad (0 \le x \le 1) \\ &= \phi_0(x - 2) \qquad (2 \le x \le 3). \end{aligned}$$

Show that ϕ is ergodic, but ϕ^2 is not. Construct a similar example of an affine transformation on $G = G_0 \oplus \{0, 1\}$. Why does this not violate Theorem 3.3.

11. In Example 5 we know that ϕ is semiergodic if a_0 separates the points on $\hat{G}_0$. Let $G_0 = \{0, 1\} = \hat{G}_0$, $a_0 = 1$, and $\gamma = (1, 1, 0, 0, 0, \ldots)$. Addition, of course, is modulo 2. Since $\tau^{*2}\gamma = \gamma$, but $\tau^*\gamma \ne \gamma$, condition (iv) of Theorem 3.3 is not satisfied. Show that $T_\phi{}^4\gamma = \gamma$, so that ϕ is not totally ergodic.

12. In Example 5b, show that (iii) of Theorem 3.3 is violated by calculating $\lambda_2(\gamma)$. Show further that

$$\begin{aligned} f(x) &= [\gamma(x)]^2 + [T_\phi \gamma(x)]^2 \\ &= \exp[\pi i(10x_1 + 12x_2 + 8x_3)/8] + \exp[\pi i(10 + 6x_1 + 4x_2 + 8x_3)/8] \end{aligned}$$

is a nontrivial invariant function for T_ϕ.

13. Let ϕ, τ, σ, a and G be as in Theorem 3.3, except we do not assume that $\hat{G}$ is torsion-free.

(a) The following implications among the conditions in Theorem 3.3 are valid:

$$\begin{array}{c} \text{(iii)} \Rightarrow \text{(ii)} \\ \Downarrow \\ \text{(iv)} \Rightarrow \text{(i)} \\ \Updownarrow \\ \text{(v)} \end{array}$$

[Hint. Show that (iv) ⇒ (i) and then that (iii) for ϕ implies (iv) for ϕ^k.]

(b) If ϕ is ergodic, then

(iii′) $\tau^{*n}(\gamma) = \gamma,\ \lambda_n(\gamma) = 1 \Rightarrow n\gamma = 0$,

and

(iv′) $\tau^{*n}(\gamma) = \gamma \Rightarrow \tau^*(n\gamma) = n\gamma$, and ϕ is semiergodic.

[Hint. To prove (iii′), define the invariant function

$$f_1(x) = \sum_{k=0}^{n-1} T_\phi{}^k\gamma(x) = \sum_{k=0}^{n-1} \lambda_k(\gamma)\tau^{*k}\gamma(x)$$

and deduce that $n = 1$ and $\gamma = 0$, or there exists a smallest integer l_1, $0 < l_1 < n$, such that $\tau^{*l_1}(\gamma) = \gamma$. Moreover, l_1 divides n, say $n = l_1 m_1$. Inductively, define integers l_p, m_p with $n = l_p m_p$, $\tau^{*l_p}(m_{p-1}\gamma) = m_{p-1}\gamma$, $\tau^{*i}(m_{p-1}\gamma) \neq m_{p-1}\gamma$ for $0 < i < l_p$ and functions f_p,

$$f_{p+1}(x) = \sum_{k=0}^{n-1} [T_\phi{}^k\gamma(x)]^{l_p} = m_p \sum_{k=0}^{l_p-1} \lambda_k(m_p\gamma)\tau^{*k}(m_p\gamma)(x)$$

and such that $0 < l_{p+1} < l_p$ or $l_p = 1$. Conclude that, for some p, we have $l_p = 1$ and $m_p\gamma = n\gamma = 0$.]

14. Let $G_0 = \{0, 1\}$ with addition modulo 2, and $G = G_0 \otimes G_0 \otimes G_0$. Define τ on G by

$$\tau(x_1, x_2, x_3) = (x_1, x_1 + x_2, x_2 + x_3),$$

and let $a = (1, 0, 0)$. Define ϕ as usual.

(a) The orbit of a under ϕ is a proper subset of G. Hence ϕ is not ergodic.

(b) If $\gamma = (0, 1, 0) \in \hat{G}$, then $\tau^{*2}(\gamma) = \gamma$, but $\tau^*(\gamma) \neq \gamma$ and $\lambda_2(\gamma) = -1$. Moreover, $\tau^{*4}(\gamma) = \gamma$ and $\lambda_4(\gamma) = 1$.

(c) Conditions (iii′) and (iv′) above are satisfied.

(d) The mapping $\psi: G \to G_0 \otimes G_0$ defined by $\psi(x_1, x_2, x_3) = (x_1, x_2)$ induces a factor $\bar{\phi}$ of ϕ on $G_0 \otimes G_0$. $\bar{\phi}$ is ergodic, but does not satisfy (iii) or (iv) of Theorem 3.3.

15. In Example 5 assume that a_0 separates the points of $\hat{G}_0$, for example, if a_0 is a topological generator of G_0. Prove that $O_\phi(0) = O_\phi(a)$ separates the points of $\hat{G}$. Is it dense in G?

Minimality

16. Suppose that τ is an ergodic automorphism and $\phi(x) = \tau(x) + a = \sigma(x) + x + a$. Show that

(a) σ^* is monic and σ is epic;

(b) ϕ is affinely isomorphic to τ under the map $\psi(x) = x + b$, where $\sigma(b) = -a$;

(c) ϕ is not minimal.

17. **(a)** Let ϕ_0 be a minimal homeomorphism of $[0, 1]$. Define ϕ on $[0, 1] \cup [2, 3]$ as in Exercise 10. Show that ϕ is minimal, but ϕ^2 is not.

(b) If X is a connected, compact metric space, and if $\phi: X \to X$ is a minimal homeomorphism, then ϕ is totally minimal.

18. Deduce from Exercise 7, Chapter II, that a uniquely ergodic affine transformation is minimal.

Quasidiscrete and Quasiperiodic Spectrum

19. Prove Proposition 3.3.

20. For any affine ϕ, show that Γ_1 consists of all those elements of $\hat{G}$ which are eigenfunctions of T_ϕ.

21. **(a)** Show that $\Gamma_n^\perp = \sigma^n(G)$, and hence that

$$\Gamma^\perp = \left(\bigcup_{n=0}^{\infty} \Gamma_n\right)^\perp = \bigcap_{n=0}^{\infty} \sigma^n(G).$$

(b) Conclude that ϕ has quasidiscrete spectrum iff $\bigcap_{n=0}^{\infty} \sigma^n(G) = \{0\}$.

22. Show that the affine transformation ϕ maps the group generated by a and $\sigma(G)$ into itself, and hence that a minimal ϕ is semiergodic. Show that the converse is false even for transformations with quasidiscrete spectrum.

23. **(a)** Let $\Phi = (X, \phi)$ be a classical dynamical system, and denote by $\mathcal{G} = \mathcal{G}(X)$ the group of complex-valued, continuous functions of constant absolute value one. Show that $\alpha(f) = f(x_0)$ is a homomorphism such as is required in the proof of Theorem 3.4.

(b) If Φ is either a classical or abstract dynamical system and if $\mathcal{G}$ and α are defined appropriately, we shall say that the pair (Φ, α) is *semiergodic* if

$$f \in \mathcal{G}, \quad \tilde{\phi}(f) = f, \quad \alpha(f) = 1 \Rightarrow f = 1.$$

Show that Theorems 3.4 and 3.5 remain valid for semiergodic systems.

24. Let ϕ be an ergodic affine transformation.

(a) Let $f \in B_1$ be an arbitrary eigenfunction of T_ϕ, say $T_\phi f = \lambda f$. Expand f in a (generalized) Fourier series

$$f = \sum_{\gamma \in \hat{G}} (f, \gamma)\gamma$$

and show as in the proof of Theorem 3.3 that

$$f = \sum_{\substack{\tau^*\gamma = \gamma \\ \gamma(a) = \gamma}} (f, \gamma)\gamma$$

Deduce that there is only one such γ for each λ, and hence that $A_1 \cong \Gamma_1$, $B_1 = \{c\gamma : c \in K, \gamma \in \Gamma_1\} \cong K \oplus \Gamma_1$.

(b) Show that $A_n \cong \Gamma_n$ and $B_n \cong K \oplus \Gamma_n$.

25. Verify the details of the proof of Theorem 3.7.

26. **(a)** Let τ be an automorphism of G, and let λ be an eigenvalue of T_τ. Show that λ is a kth root of unity for some k, and the eigenfunctions associated with λ are of the form

$$f = \sum_{\tau^{*k}\gamma=\gamma} (f, \gamma)\gamma.$$

(b) The linear space in $L_2(X)$ spanned by Δ_1 includes all the eigenfunctions of T_τ.

(c) If $\phi(x) = \tau(x) + a$, the eigenfunctions of T_ϕ corresponding to the eigenvalue λ are all of the form

$$f = \sum_{k=1}^{\infty} \sum_{\substack{\tau^{*k}\gamma = \gamma \\ \lambda^k = \lambda_k(\gamma)}} (f, \gamma)\gamma.$$

Ergodic Automorphisms

27. Show that Proposition 3.7 remains true if τ is only assumed to be a continuous affine transformation of the compact metric group G.

28. Use the facts that the annihilator of $O_\tau(a)$ and the annihilator of the group Gp $\text{Orb}_\tau(a)$ generated by $O_\tau(a)$ are the same and that $H^{\perp\perp} = \overline{H}$, the closure of H, for any group $H \subseteq G$ to prove Proposition 3.8.

29. **(a)** If $O_\tau(a)$ separates the points of $\hat{G}$, show that $\phi(x) = \tau(x) + a$ is semiergodic.

(b) Show that the ϕ of Exercise 14 is semiergodic.

30. **(a)** Let us say that the system (G, τ), where τ is a continuous automorphism, is *weakly topologically ergodic* if $A \subseteq G$, $\mathscr{E}(\tau)A \subseteq A$ implies A is dense or A is nowhere dense in G. If G is metrizable and (G, τ) is weakly topologically ergodic, show that (G, τ) is monothetic.

(b) If (G, I) is weakly topologically ergodic, then (G, τ) is weakly topologically ergodic for any τ.

(c) (K^n, I) and (K^ω, I) are weakly topologically ergodic. Hence (K^n, τ) and (K^ω, τ) are monothetic for any τ.

31. If G is a monothetic group, show that (G, τ) is monothetic for any automorphism τ. Which of the groups K^n, K^ω, $\hat{K}_d$, $\hat{K}_d{}^n$, $\hat{K}_d{}^\omega$ are monothetic? Which are algebraically monothetic?

32. *Algebraically monothetic groups.* Let $T(H)$ denote the torsion subgroup of the group H.

(a) Show by transfinite induction on the elements of infinite order that any monomorphism of $T(H)$ onto a shift-invariant subgroup of K_d^{ω} can be extended to a monomorphism of H onto a shift-invariant subgroup of K_d^{ω}, provided that the cardinality of H does not exceed the cardinality c of the continuum.

(b) Theorem *A compact abelian group G is algebraically monothetic iff the cardinality of $\hat{G}$ is no greater than c and $T(\hat{G})$ is isomorphic to a shift-invariant subgroup of K_d^{ω}.*

(c) Every separable (and hence every metrizable) connected compact abelian group is algebraically monothetic (in fact monothetic).

Note The next two exercises require some knowledge of abstract group theory. See, for example, Kaplansky [38]. In addition, the theorem in Exercise 32(b) is used.

33. (a) If $\hat{G} = \bigoplus_{j=1}^{s} Z(p^{k_j})$, where p is prime and $1 \leq k_1 \leq k_2 \leq \cdots \leq k_s \leq \infty$, and if τ^* is defined on $\hat{G}$ by

$$\tau^*(t_1, \ldots, t_s) = (t_1, t_1 + t_2, t_2 + t_3, \ldots, t_{s-1} + t_s),$$

then (G, τ) is monothetic.

(b) If $\hat{G}$ is finitely generated, then G is algebraically monothetic.

34. If G is separable and $\hat{G}$ is divisible, then G is algebraically monothetic.

Quasidiscrete Spectrum II

35. Show that a totally ergodic abstract or totally minimal classical dynamical system can have no eigenvalues λ with $\lambda^n = 1$.

36. Complete the proof of Theorem 3.9(ii) for totally minimal classical systems.

CHAPTER

IV

Entropy

1. CONDITIONAL EXPECTATION AND KOLMOGOROV ENTROPY

In Chapter I we introduced the notion of equivalence of two abstract dynamical systems $\Phi_1 = (X_1, \mathscr{B}_1, \mu_1, \phi_1)$ and $\Phi_2 = (X_2, \mathscr{B}_2, \mu_2, \phi_2)$. Thus Φ_1 and Φ_2 are *equivalent* if there exists an isomorphism $\psi^*: \mathscr{B}_2 \to \mathscr{B}_1$ of the measure algebras $\mathscr{B}_2$ and $\mathscr{B}_1$ satisfying $\phi_1^{-1}\psi^* = \psi^*\phi_2^{-1}$ modulo sets of measure zero (Definition 1.2). In most cases, we may assume that $\psi^* = \psi^{-1}$ (mod 0), where $\psi: X_1 \to X_2$ is an invertible measure-preserving transformation. In order to distinguish this notion from other types of equivalence or isomorphism for dynamical systems, let us agree to write $\Phi_1 \overset{\text{m}}{=} \Phi_2$ when Φ_1 and Φ_2 are equivalent in the above sense. We shall say that Φ_1 and Φ_2 are *m-isomorphic*.

We turn now to a study of the "isomorphism problem" for abstract dynamical systems. When are two dynamical systems m-isomorphic? A complete solution of this problem should include two things: a useful set of criteria for determining when two dynamical systems are isomorphic and the concrete realization of a representative system from each isomorphism class. For the subclass of dynamical systems with quasidiscrete spectrum, Theorems 3.9(ii) and 3.5 provide such a solution. Two systems are isomorphic iff they have "equivalent" sets of quasieigenvalues, and $\tilde{\Phi}$ can be realized as a factor of $\hat{\Phi}$. On the other hand, the general problem is far from solution.

A big step in the solution of this problem was made in 1958 when Kolmogorov [41] introduced the definition (later modified slightly by Sinaĭ [58]) of the entropy of an abstract dynamical system and showed that it was an isomorphism invariant. The next big step occurred in 1970 when D. S. Ornstein showed that entropy is a "complete" invariant for invertible Bernoulli shifts (Example 3 of Chapter I). That is, two Bernoulli shifts are m-isomorphic iff they have the same entropy. We shall discuss this result and related matters in Chapter V.

Other isomorphism invariants need to be mentioned at this time. For example, ergodicity and mixing, as well as invertibility, are easily seen to be preserved by m-isomorphism.

Associated with each abstract dynamical system Φ we have a linear operator T_ϕ on the Hilbert space $L_2(X)$ and an affine transformation $\tilde{\phi}$ on the compact group $G = G(\Phi)$ where $\hat{G} = \sigma^*(\mathscr{G})$ and where $\sigma^* = \sigma_\Phi{}^*$ and $\mathscr{G} = \mathscr{G}(\Phi)$ are defined in Section III.6. It is easily established (Exercise 1) that $\Phi_1 \overset{\text{m}}{=} \Phi_2$ implies that T_{ϕ_1} and T_{ϕ_2} are isomorphic as operators on Hilbert space, and that $\tilde{\phi}_1$ and $\tilde{\phi}_2$ are isomorphic as affine transformations. Thus the adjunct transformations T_ϕ and $\tilde{\phi}$ provide an abundance of isomorphism invariants of the dynamical system Φ.

As shown in Section I.3, all Bernoulli shifts have countable Lebesgue spectrum. It follows that they are strongly mixing and induce isomorphic operators T_ϕ. The first important result using Kolmogorov entropy was that, for example, the shift on two points is *not* m-isomorphic to the shift on three points.

In order to define the Kolmogorov entropy $h(\Phi)$ of Φ, we must digress briefly to discuss *conditional expectation operators* on $L_2(X)$.

Definition 4.1 Let $(X, \mathscr{B}, \mu)$ be a normalized measure space $(\mu(X) = 1)$, and let $\mathscr{C} \subseteq \mathscr{B}$ be a sub-σ-algebra. Let E denote the embedding operator. $Ef = f$ of $L_2(X, \mathscr{C}, \mu)$ into $L_2(X, \mathscr{B}, \mu)$, and let E^* be the adjoint of E. The *conditional expectation operator* $\mathbb{E}_\mathscr{C}$ is defined by

$$\mathbb{E}_\mathscr{C} = EE^*. \tag{1}$$

The operator $\mathbb{E}_\mathscr{C}$ acts on $L_2(X, \mathscr{B}, \mu)$ and leaves invariant the subspace $L_2(X, \mathscr{C}, \mu)$. For each $f \in L_2(X, \mathscr{B}, \mu)$, we say that $\mathbb{E}_\mathscr{C} f$ is the *conditional expectation of f given $\mathscr{C}$*.

Proposition 4.1 *The operator $\mathbb{E}_\mathscr{C}$ is doubly stochastic and self-adjoint.*

Proof Obvious.

Some other useful properties of $\mathbb{E}_{\mathscr{C}}$ are summarized in the following proposition. Note that, according to Proposition 4.1, $\mathbb{E}_{\mathscr{C}}$ may be assumed to act also on $L_1(X, \mathscr{B}, \mu)$.

Proposition 4.2 *For all functions f, $g \in L_1(X, \mathscr{B}, \mu)$ and all σ-algebras $\mathscr{C}, \mathscr{C}_1, \mathscr{C}_2 \subseteq \mathscr{B}$ the following are valid:*

(i) $\mathbb{E}_{\mathscr{C}}{}^2 = \mathbb{E}_{\mathscr{C}}$.
(ii) *If f is $\mathscr{C}$-measurable, and $f \cdot g \in L_1$, then*

$$\mathbb{E}_{\mathscr{C}}(f \cdot g) = f \cdot \mathbb{E}_{\mathscr{C}}\, g. \tag{2}$$

(iii) $\mathscr{C}_1 \subseteq \mathscr{C}_2 \Rightarrow \mathbb{E}_{\mathscr{C}_1}\mathbb{E}_{\mathscr{C}_2} = \mathbb{E}_{\mathscr{C}_2}\mathbb{E}_{\mathscr{C}_1} = \mathbb{E}_{\mathscr{C}_1}$.
(iv) $\mathbb{E}_{\mathscr{B}}\, f = f$.
(v) $\mathbb{E}_{\mathscr{N}}\, f = \int_X f\, d\mu$, *where* $\mathscr{N} = \{\phi, X\}$.
(vi) $\int_C \mathbb{E}_{\mathscr{C}}\, f\, d\mu = \int_C f\, d\mu$ *for all* $C \in \mathscr{C}$.
(vii) $\mathbb{E}_{\mathscr{C}}\, f$ *is $\mathscr{C}$-measurable.*
(viii) *If g has properties* (vi) *and* (vii) *of $\mathbb{E}_{\mathscr{C}}\, f$, then $g = \mathbb{E}_{\mathscr{C}}\, f$ μ-almost everywhere.*

Proof Property (i) follows from the equality $E^*E = I$, the identity on $L_2(X, \mathscr{C}, \mu)$, which is equivalent to $(Ef, Ef) = \int_X |f|^2\, d\mu = (f, f)$.

In (ii) we assume f is $\mathscr{C}$-measurable, so that as functions $f = Ef$. Thus (2) is equivalent to

$$\mathbb{E}_{\mathscr{C}}(Ef \cdot g) = Ef \cdot \mathbb{E}_{\mathscr{C}}\, g \tag{2'}$$

for $f \in L_2(X, \mathscr{C}, \mu)$. Let $h \in L_\infty(X, \mathscr{B}, \mu)$. Then, since E is multiplicative and $\mathbb{E}_{\mathscr{C}}$ is self-adjoint,

$$\begin{aligned}
(\mathbb{E}_{\mathscr{C}}(Ef \cdot g), h) &= (Ef \cdot g, \mathbb{E}_{\mathscr{C}}\, h) \\
&= (g, \overline{Ef} \cdot EE^*h) = (g, E(\bar{f} \cdot E^*h)) \\
&= (E^*g, \bar{f} \cdot E^*h) = (f \cdot E^*g, E^*h) \\
&= (Ef \cdot EE^*g, h) = (Ef \cdot \mathbb{E}_{\mathscr{C}}\, g, h),
\end{aligned}$$

from which follows (2′).

(iii) Let E_i $(i = 1, 2)$ be the embedding operator of $L_2(X, \mathscr{C}_i, \mu)$ into $L_2(X, \mathscr{B}, \mu)$, and let E_0 be the embedding of $L_2(X, \mathscr{C}_1, \mu)$ into $L_2(X, \mathscr{C}_2, \mu)$. Then $E_1 = E_2 E_0$, and so

$$\mathbb{E}_{\mathscr{C}_1} = E_1 E_1{}^* = E_2 E_0 E_0{}^* E_2{}^*.$$

Thus

$$\mathbb{E}_{\mathscr{C}_1}\mathbb{E}_{\mathscr{C}_2} = E_2 E_0 E_0{}^*(E_2{}^*E_2)E_2{}^* = \mathbb{E}_{\mathscr{C}_1},$$

$$\mathbb{E}_{\mathscr{C}_2}\mathbb{E}_{\mathscr{C}_1} = E_2(E_2{}^*E_2)E_0 E_0{}^*E_2{}^* = \mathbb{E}_{\mathscr{C}_1},$$

since $E_2{}^*E_2 = I$.

(iv) This is trivial, since $E = I$.

(v) For $\mathscr{C} = \mathscr{N}$, E^*f is a constant. Since $E1 = 1$, that constant must be $\int f\, d\mu$.

The last three statements are proved in the exercises at the end of this chapter. ▮

Certain properties of the conditional expectation needed in the sequel depend only on the fact that $\mathbb{E}_{\mathscr{C}}$ is doubly stochastic. Two such properties are given in the following propositions.

Proposition 4.3 *Suppose $f_n, g \in L_p(X, \mathscr{B}, \mu)$. $|f_n| \le g$ and $f_n \to f$ a.e. Then, for each doubly stochastic operator T, $Tf_n \to Tf$ a.e. and in the mean of order p $(1 \le p \le \infty)$.*

Proof The L_p convergence follows immediately from the dominated convergence theorem and the inequality

$$\|Tf_n - Tf\|_p \le \|f_n - f\|_p.$$

Almost everywhere convergence for a monotone sequence of functions follows from positivity of the operator T, the equality $\int Tf\, d\mu = \int f\, d\mu$, and the monotone convergence theorem. Dominated convergence then follows from monotone convergence in the usual way. ▮

A real-valued function F of a real variable is said to be *convex* if

$$F\left(\sum_{k=1}^{n} t_k x_k\right) \le \sum_{k=1}^{n} t_k F(x_k) \tag{3}$$

for all choices of x_k in the domain of F and all $t_k \ge 0$ such that $\sum_{k=1}^{n} t_k = 1$.

Proposition 4.4 (*Jensen's inequality*) *Let $F\colon [a, b] \to R$ be a continuous, convex function, where $0 \le a < b < \infty$ and R denotes the real numbers. Let T be a doubly stochastic operator. Then*

$$F \circ Tf \le T(F \circ f) \tag{4}$$

for each $f \in L_1(X, \mathscr{B}, \mu)$ with $f(X) \subseteq [a, b]$.

Proof Suppose first that f is a simple function:

$$f = \sum_{k=1}^{n} c_k \chi_{B_k},$$

where the $B_k \in \mathscr{B}$ are pairwise disjoint with union X. Then

$$\sum_{k=1}^{n} T\chi_{B_k} = T1 = 1, \text{ and } F \circ f = \sum_{k=1}^{n} F(c_k)\chi_{B_k},$$

so that by (3)

$$T(F \circ f) = \sum_{k=1}^{n} F(c_k) T\chi_{B_k} \geq F \circ \sum_{k=1}^{n} c_k T\chi_{B_k} = F \circ Tf \qquad \text{a.e.}$$

In the general case, let f_n $(n = 1, 2, \ldots)$ be simple functions with $0 \leq f_n \uparrow f$. Then $Tf_n \uparrow Tf$ a.e. (by Proposition 4.3), so that $F \circ Tf_n \to F \circ Tf$ a.e. (by continuity of F). Moreover, $F \circ f_n \to F \circ f$ and $|F \circ f_n| \leq \max_{a \leq x \leq b} |F(x)| = c$. Again using Proposition 4.3, $T(F \circ f_n) \to T(F \circ f)$ a.e. Since (4) holds for each f_n, it also holds in the limit. ∎

Remark We shall only need Proposition 4.4 in the case $T = \mathbb{E}_{\mathscr{C}}$ and

$$F(t) = \begin{cases} t \log t & (0 < t \leq 1) \\ 0 & (t = 0). \end{cases} \tag{5}$$

The function F is used in the Kolmogorov definition of entropy as follows.

Definition 4.2 Let $\Phi = (X, \mathscr{B}, \mu, \phi)$ be a dynamical system. For each finite algebra $\mathscr{A} \subseteq \mathscr{B}$, let $\dot{\mathscr{A}}$ denote the collection of atoms of $\mathscr{A}$ (sets in $\mathscr{A}$ having positive measure but no $\mathscr{A}$-measurable subsets of positive measure). The *entropy* of $\mathscr{A}$ is

$$H(\mathscr{A}) = -\sum_{A \in \dot{\mathscr{A}}} \mu(A) \log \mu(A). \tag{6}$$

The *entropy of ϕ on $\mathscr{A}$* is

$$h(\phi, \mathscr{A}) = \overline{\lim_{n \to \infty}} \frac{1}{n} H\left(\bigvee_{i=0}^{n-1} \phi^{-i}\mathscr{A}\right), \tag{7}$$

and the *entropy of Φ* is

$$h(\Phi) = \sup_{\mathscr{A}} h(\phi, \mathscr{A}). \tag{8}$$

The sup in (8) is over all finite algebras $\mathscr{A} \subseteq \mathscr{B}$. The algebra $\bigvee_{i=0}^{n-1} \phi^{-i}\mathscr{A}$ in (7) is the smallest algebra containing each of the algebras

$$\phi^{-i}\mathscr{A} = \{\phi^{-i}(A) : A \in \mathscr{A}\}. \tag{9}$$

In the following sections we shall show that $h(\Phi)$ is an m-isomorphism invariant, that $h(\phi, \mathscr{A})$ is finite for each finite algebra $\mathscr{A}$, that the $\overline{\lim}$ in (7) is actually a lim, and that under certain conditions the sup in (8) is actually attained, so that $h(\Phi) < \infty$.

2. THE INFORMATION FUNCTION AND FINITENESS OF $h(\phi, \mathscr{A})$

In order to study the function $H(\mathscr{A})$ defined by Eq. (6), we introduce a related function, called the *information function*. Thus for each finite algebra $\mathscr{A}$ and each $x \in X$ we define $I(\mathscr{A}, x)$ by

$$I(\mathscr{A}, x) = - \sum_{A \in \mathscr{A}} \chi_A(x) \log \mu(A), \tag{10}$$

so that

$$H(\mathscr{A}) = \int_X I(\mathscr{A}, x)\, \mu(dx). \tag{11}$$

Similarly, we define the *conditional information function* and *conditional entropy* of $\mathscr{A}$ given $\mathscr{C}$ by

$$I(\mathscr{A} | \mathscr{C}, x) = - \sum_{A \in \mathscr{A}} \chi_A(x) \log \mathbb{E}_{\mathscr{C}} \chi_A(x), \tag{12}$$

and

$$H(\mathscr{A} | \mathscr{C}) = \int_X I(\mathscr{A} | \mathscr{C}, x)\, \mu(dx). \tag{13}$$

Note that $\mathbb{E}_{\mathscr{C}} \chi_A(x) = \mu(A)$ if $\mathscr{C} = \mathscr{N}$, so that $I(\mathscr{A} | \mathscr{N}, x) = I(\mathscr{A}, x)$ and $H(\mathscr{A} | \mathscr{N}) = H(\mathscr{A})$. We may think of $\mathbb{E}_{\mathscr{C}} \chi_A$ as the *conditional measure* of A given $\mathscr{C}$.

Let $f \in L_1(X, \mathscr{B}, \mu)$. Then $T_\phi \mathbb{E}_{\mathscr{C}} f \in L_1(X, \phi^{-1}\mathscr{C}, \mu)$. Moreover, for any $B \in \mathscr{C}$ we have

$$\begin{aligned} \int_{\phi^{-1}B} T_\phi \mathbb{E}_{\mathscr{C}} f\, d\mu &= \int_{\phi^{-1}B} (\mathbb{E}_{\mathscr{C}} f) \circ \phi\, d\mu \\ &= \int_B \mathbb{E}_{\mathscr{C}} f\, d\mu = \int_B f\, d\mu = \int_{\phi^{-1}B} f \circ \phi\, d\mu. \end{aligned}$$

According to Proposition 4.2(viii),

$$T_\phi \mathbb{E}_{\mathscr{C}} f = \mathbb{E}_{\phi^{-1}\mathscr{C}}(f \circ \phi) = \mathbb{E}_{\phi^{-1}\mathscr{C}}(T_\phi f). \tag{14}$$

Letting $f = \chi_A$ in (14) yields

$$\mathbb{E}_{\phi^{-1}\mathscr{C}} \chi_{\phi^{-1}A} = T_\phi \mathbb{E}_{\mathscr{C}} \chi_A .$$

Hence

$$I(\phi^{-1}\mathscr{A} | \phi^{-1}\mathscr{C}, x) = - \sum_{A \in \mathscr{A}} \chi_A(\phi x) \log \mathbb{E}_{\mathscr{C}} \chi_A(\phi x) = I(\mathscr{A} | \mathscr{C}, \phi x). \tag{15}$$

Now let $\mathscr{A}_1$, $\mathscr{A}_2$ be finite subalgebras of $\mathscr{B}$, and let $\mathscr{C} \subseteq \mathscr{B}$ be an arbitrary σ-algebra. The *join* of $\mathscr{A}_1$ and $\mathscr{A}_2$, denoted by $\mathscr{A}_1 \vee \mathscr{A}_2$, is the smallest σ-algebra containing $\mathscr{A}_1$ and $\mathscr{A}_2$. Its atoms are exactly the intersections $A \cap B$ of positive measure with $A \in \mathscr{A}_1$ and $B \in \mathscr{A}_2$. We shall show that

$$I(\mathscr{A}_1 \vee \mathscr{A}_2 | \mathscr{C}, x) = I(\mathscr{A}_1 | \mathscr{C}, x) + I(\mathscr{A}_2 | \mathscr{A}_1 \vee \mathscr{C}, x). \tag{16}$$

Suppose $A \in \mathscr{A}_1$, $B \in \mathscr{B}$, $C \in \mathscr{C}$. Then A is the disjoint union of those atoms F of $\mathscr{A}_1$ such that $A \cap F \neq \varnothing$. From this and parts (ii) and (vi) of Proposition 4.2, we see that the integral

$$\int_{A \cap C} \left[\sum_{F \in \mathscr{A}_1} \chi_F(x) \frac{\mathbb{E}_{\mathscr{C}} \chi_{B \cap F}(x)}{\mathbb{E}_{\mathscr{C}} \chi_F(x)} \right] \mu(dx),$$

where the fraction is taken to be zero when $\mathbb{E}_{\mathscr{C}} \chi_F(x) = 0$, is equal to

$$\begin{aligned}
&\sum_{F \in \mathscr{A}_1} \int_C \chi_{A \cap F}(x) \frac{\mathbb{E}_{\mathscr{C}} \chi_{B \cap F}(x)}{\mathbb{E}_{\mathscr{C}} \chi_F(x)} \mu(dx) \\
&\quad = \sum_{F \in \mathscr{A}_1, A \cap F \neq \varnothing} \int_C \chi_F(x) \frac{\mathbb{E}_{\mathscr{C}} \chi_{B \cap F}(x)}{\mathbb{E}_{\mathscr{C}} \chi_F(x)} \mu(dx) \\
&\quad = \sum_{F \in \mathscr{A}_1, A \cap F \neq \varnothing} \int_C \mathbb{E}_{\mathscr{C}} \left[\chi_F(x) \frac{\mathbb{E}_{\mathscr{C}} \chi_{B \cap F}(x)}{\mathbb{E}_{\mathscr{C}} \chi_F(x)} \right] \mu(dx) \\
&\quad = \sum_{F \in \mathscr{A}_1, A \cap F \neq \varnothing} \int_C \mathbb{E}_{\mathscr{C}} \chi_F(x) \frac{\mathbb{E}_{\mathscr{C}} \chi_{B \cap F}(x)}{\mathbb{E}_{\mathscr{C}} \chi_F(x)} \mu(dx) \\
&\quad = \sum_{F \in \mathscr{A}_1, A \cap F \neq \varnothing} \mu(B \cap F \cap C) = \mu(B \cap A \cap C).
\end{aligned}$$

Since finite unions of intersections of the form $A \cap C$ are dense in the σ-algebra $\mathscr{A}_1 \vee \mathscr{C}$, it follows that for any $D \in \mathscr{A}_1 \vee \mathscr{C}$ we have

$$\int_D \sum_{F \in \mathscr{A}_1} \chi_F(x) \frac{\mathbb{E}_{\mathscr{C}} \chi_{B \cap F}(x)}{\mathbb{E}_{\mathscr{C}} \chi_F(x)} \mu(dx) = \mu(B \cap D).$$

From Proposition 4.2(viii), it follows that

$$\mathbb{E}_{\mathscr{A}_1 \vee \mathscr{C}} \chi_B(x) = \sum_{F \in \mathscr{A}_1} \chi_F(x) \frac{\mathbb{E}_{\mathscr{C}} \chi_{B \cap F}(x)}{\mathbb{E}_{\mathscr{C}} \chi_F(x)} \qquad \text{a.e.} \tag{17}$$

for each $B \in \mathscr{B}$.

Now let us evaluate the left side of Eq. (16), making use of (17). From (12) we have

$$\begin{aligned}
I(\mathscr{A}_1 \vee \mathscr{A}_2 | \mathscr{C}, x) &= - \sum_{A \in \mathscr{A}_1} \sum_{B \in \mathscr{A}_2} \chi_{A \cap B}(x) \log \mathbb{E}_{\mathscr{C}} \chi_{A \cap B}(x) \\
&= - \sum_{A \in \mathscr{A}_1} \sum_{B \in \mathscr{A}_2} \chi_A(x)\, \chi_B(x) \log \left[\mathbb{E}_{\mathscr{C}} \chi_A(x) \frac{\mathbb{E}_{\mathscr{C}} \chi_{A \cap B}(x)}{\mathbb{E}_{\mathscr{C}} \chi_A(x)} \right] \\
&= - \sum_{A \in \mathscr{A}_1} \sum_{B \in \mathscr{A}_2} \chi_A(x)\, \chi_B(x) \log \mathbb{E}_{\mathscr{C}} \chi_A(x) \\
&\quad - \sum_{A \in \mathscr{A}_1} \sum_{B \in \mathscr{A}_2} \chi_A(x)\, \chi_B(x) \log \frac{\mathbb{E}_{\mathscr{C}} \chi_{A \cap F}(x)}{\mathbb{E}_{\mathscr{C}} \chi_A(x)} \\
&= - \sum_{B \in \mathscr{A}_2} \chi_B(x) \left[\sum_{A \in \mathscr{A}_1} \chi_A(x) \log \mathbb{E}_{\mathscr{C}} \chi_A(x) \right] \\
&\quad - \sum_{B \in \mathscr{A}_2} \chi_B(x) \log \left[\sum_{A \in \mathscr{A}_1} \chi_A(x) \frac{\mathbb{E}_{\mathscr{C}} \chi_{A \cap B}(x)}{\mathbb{E}_{\mathscr{C}} \chi_A(x)} \right] \\
&= \sum_{B \in \mathscr{A}_2} \chi_B(x)\, I(\mathscr{A}_1 | \mathscr{C}, x) - \sum_{B \in \mathscr{A}_2} \chi_B(x) \log \mathbb{E}_{\mathscr{A}_1 \vee \mathscr{C}} \chi_B(x) \\
&= I(\mathscr{A}_1 | \mathscr{C}, x) + I(\mathscr{A}_2 | \mathscr{A}_1 \vee \mathscr{C}, x),
\end{aligned}$$

as asserted.

Equation (16) expresses a type of additivity for the information function. Let us see how this extends to finite joins of finite algebras and integrate to get a formula for the entropy $H(\bigvee_{i=0}^{n-1} \phi^{-i}\mathscr{A})$ occurring in (7).

Let $\mathscr{N} = \mathscr{A}_0, \mathscr{A}_1, \ldots, \mathscr{A}_{n+1} \subseteq \mathscr{B}$ be finite algebras. Setting $\mathscr{C} = \mathscr{N}$ in (16) gives

$$I\left(\bigvee_{i=0}^{k} \mathscr{A}_i \right) = I\left(\bigvee_{i=0}^{k-1} \mathscr{A}_i \right) + I\left(\mathscr{A}_k \,\middle|\, \bigvee_{i=0}^{k-1} \mathscr{A}_i \right),$$

for each $k = 1, 2, \ldots, n+1$. Noting that $I(\mathscr{A}_0) = I(\mathscr{N}) = 0$, we obtain by adding on k

$$I\left(\bigvee_{i=0}^{n+1} \mathscr{A}_i \right) = \sum_{k=1}^{n+1} I\left(\mathscr{A}_k \,\middle|\, \bigvee_{i=0}^{k-1} \mathscr{A}_i \right). \tag{18}$$

Proposition 4.5 *For any finite algebra $\mathscr{A} \subseteq \mathscr{B}$ and each $n = 1, 2, \ldots,$ we have*

$$I\left(\bigvee_{i=0}^{n} \phi^{-i}\mathscr{A}, x\right) = I(\mathscr{A}, \phi^n x) + \sum_{k=1}^{n} I\left(\mathscr{A} \,\middle|\, \bigvee_{i=1}^{k} \phi^{-i}\mathscr{A}, \phi^{n-k}x\right), \tag{19}$$

and hence

$$H\left(\bigvee_{i=0}^{n} \phi^{-i}\mathscr{A}\right) = H(\mathscr{A}) + \sum_{k=1}^{n} H\left(\mathscr{A} \,\middle|\, \bigvee_{i=1}^{k} \phi^{-i}\mathscr{A}\right). \tag{20}$$

Proof We apply (18) to $\mathscr{A}_i = \phi^{-(n+1-i)}\mathscr{A}$ $(i = 1, 2, \ldots, n+1)$ to obtain

$$I\left(\bigvee_{i=0}^{n} \phi^{-i}\mathscr{A}\right) = I\left(\bigvee_{i=1}^{n+1} \phi^{-(n+1-i)}\mathscr{A}\right) = I\left(\bigvee_{i=0}^{n+1} \mathscr{A}_i\right) = \sum_{k=1}^{n+1} I\left(\mathscr{A}_k \,\middle|\, \bigvee_{i=0}^{k-1} \mathscr{A}_i\right).$$

Again recalling that $\mathscr{A}_0 = \mathscr{N}$, so that $\mathscr{A}_0 \vee \mathscr{C} = \mathscr{C}$ for any $\mathscr{C}$, we see from (15) that

$$\begin{aligned} I\left(\bigvee_{i=0}^{n} \phi^{-i}\mathscr{A}_1, x\right) &= I(\mathscr{A}_1, x) + \sum_{k=2}^{n+1} I\left(\mathscr{A}_k \,\middle|\, \bigvee_{i=1}^{k-1} \mathscr{A}_i, x\right) \\ &= I(\phi^{-n}\mathscr{A}, x) + \sum_{k=1}^{n} I\left(\phi^{-(n-k)}\mathscr{A} \,\middle|\, \bigvee_{i=1}^{k} \phi^{-(n+1-i)}\mathscr{A}, x\right) \\ &= I(\mathscr{A}, \phi^n x) + \sum_{k=1}^{n} I\left(\mathscr{A} \,\middle|\, \bigvee_{i=1}^{k} \phi^{-(k+1-i)}\mathscr{A}, \phi^{n-k}x\right), \end{aligned}$$

which is clearly the same as (19). Equation (20) now follows by (11) and the fact ϕ is measure preserving. ∎

Next let us introduce the function $J(\mathscr{A} \mid \mathscr{C}, x) = \mathbb{E}_{\mathscr{C}} I(\mathscr{A} \mid \mathscr{C}, x)$. According to Proposition 4.2(ii) and linearity of $\mathbb{E}_{\mathscr{C}}$, we have

$$J(\mathscr{A} \mid \mathscr{C}, x) = -\sum_{A \in \mathscr{A}} \mathbb{E}_{\mathscr{C}} \chi_A(x) \log \mathbb{E}_{\mathscr{C}} \chi_A(x).$$

From Proposition 4.2(vi) and (13), we also have

$$H(\mathscr{A} \mid \mathscr{C}) = \int_X J(\mathscr{A} \mid \mathscr{C}, x)\, \mu(dx). \tag{21}$$

Proposition 4.6 *Suppose $\mathscr{C}_1 \subseteq \mathscr{C}_2 \subseteq \mathscr{B}$ are σ-algebras, and $\mathscr{A} \subseteq \mathscr{B}$ is a finite algebra. Then*

$$H(\mathscr{A} \mid \mathscr{C}_2) \leq H(\mathscr{A} \mid \mathscr{C}_1). \tag{22}$$

Proof We shall apply Jensen's inequality, Proposition 4.4, to the function F defined by Eq. (5) with $T = \mathbb{E}_{\mathscr{C}_1}$ and $f = \mathbb{E}_{\mathscr{C}_2}\chi_A$ for each $A \in \dot{\mathscr{A}}$. According to Proposition 4.2(iii), we have

$$Tf = \mathbb{E}_{\mathscr{C}_1}\mathbb{E}_{\mathscr{C}_2}\chi_A = \mathbb{E}_{\mathscr{C}_1}\chi_A.$$

Thus

$$F \circ Tf = \mathbb{E}_{\mathscr{C}_1}\chi_A \log \mathbb{E}_{\mathscr{C}_1}\chi_A \leq T(F \circ f) = \mathbb{E}_{\mathscr{C}_1}[\mathbb{E}_{\mathscr{C}_2}\chi_A \log \mathbb{E}_{\mathscr{C}_2}\chi_A].$$

Multiplying by -1 and adding on $A \in \dot{\mathscr{A}}$ gives by (20)

$$J(\mathscr{A}|\mathscr{C}_1) \geq \mathbb{E}_{\mathscr{C}_1} J(\mathscr{A}|\mathscr{C}_2).$$

Integrating and recalling (21) gives (22). ∎

Theorem 4.1 *For any finite algebra $\mathscr{A} \subseteq \mathscr{B}$ the limit*

$$h(\phi, \mathscr{A}) = \lim_{n\to\infty} \frac{1}{n} H\left(\bigvee_{i=0}^{n-1} \phi^{-i}\mathscr{A}\right) = \lim_{n\to\infty} H\left(\mathscr{A}\,\middle|\,\bigvee_{i=1}^{n} \phi^{-i}\mathscr{A}\right) \tag{23}$$

exists and is finite.

Proof By Proposition 4.6 the sequence $H(\mathscr{A}|\bigvee_{i=1}^{n} \phi^{-i}\mathscr{A})$ is nonincreasing, hence has a finite nonnegative limit as $n \to \infty$. The equality of the two limits in (23) follows from Eq. (20) and the fact that a convergent sequence is Cesaro summable to the same limit.

Corollary 4.1.1 $h(\phi, \mathscr{A}) \leq H(\mathscr{A})$.

We conclude this section by deducing monotonicity properties of the functions $H(\mathscr{A}|\mathscr{C})$ and $h(\mathscr{A}, \phi)$ and showing that the entropy $h(\Phi)$ is an isomorphism invariant.

Proposition 4.7 *Let $\mathscr{A}_1 \subseteq \mathscr{A}_2$ be finite subalgebras of $\mathscr{B}$ and let $\mathscr{C} \subseteq \mathscr{B}$ be an arbitrary σ-algebra. Then*

$$I(\mathscr{A}_1|\mathscr{C}) \leq I(\mathscr{A}_2|\mathscr{C}), \tag{24}$$

$$H(\mathscr{A}_1|\mathscr{C}) \leq H(\mathscr{A}_2|\mathscr{C}), \tag{25}$$

and

$$h(\phi, \mathscr{A}_1) \leq h(\phi, \mathscr{A}_2). \tag{26}$$

Proof (24) follows from (16). The remainder of the proof is direct and is left to the exercises. ∎

Remark Inequality (26) is the motivation for definition (8) of $h(\Phi)$.

Theorem 4.2 *Suppose* $\Phi_1 | \Phi_2$. *Then* $h(\Phi_1) \leq h(\Phi_2)$. *In particular,* $h(\Phi)$ *is an m-isomorphism invariant.*

Proof Suppose $\psi\colon \Phi_2 \to \Phi_1$. Clearly, we have for each finite subalgebra $\mathscr{A} \subseteq \mathscr{B}_1$

$$H(\psi^{-1}\mathscr{A}) = H(\mathscr{A})$$

for μ_2-almost all $x \in X_2$. Thus

$$\begin{aligned} h(\phi_1, \mathscr{A}) = \lim_{n\to\infty} \frac{1}{n} H\left(\bigvee_{i=0}^{n-1} \phi_1^{-i}\mathscr{A}\right) &= \lim_{n\to\infty} \frac{1}{n} H\left(\psi^{-1}\left(\bigvee_{i=0}^{n-1} \phi_1^{-i}\mathscr{A}\right)\right) \\ &= \lim_{n\to\infty} \frac{1}{n} H\left(\bigvee_{i=0}^{n-1} \psi^{-1}\phi_1^{-i}\mathscr{A}\right) \\ &= \lim_{n\to\infty} \frac{1}{n} H\left(\bigvee_{i=0}^{n-1} \phi_2^{-i}\psi^{-1}\mathscr{A}\right) \\ &= h(\phi_2, \psi^{-1}\mathscr{A}). \end{aligned}$$

As $\mathscr{A}$ ranges over the finite subalgebras of $\mathscr{B}_1$, $\psi^{-1}\mathscr{A}$ ranges over a subset of the finite subalgebras of $\mathscr{B}_2$. Hence $h(\Phi_1) \leq h(\Phi_2)$, as asserted. ∎

3. SINAĬ'S THEOREM AND GENERATORS

In this section we prove a very powerful theorem due to Sinaĭ [58]. Sinaĭ's theorem (Theorem 4.3) enables us, in many interesting cases, to evaluate $h(\Phi) = h(\phi, \mathscr{A})$ for certain finite algebras $\mathscr{A}$. That is, Sinaĭ's theorem tells us when the sup in Eq. (7) is attained. First of all, though, we need to prove a weak form of the martingale theorem (Lemma 4.1) and deduce an alternate formula for $h(\phi, \mathscr{A})$.

Lemma 4.1 *Let* $\mathscr{C}_n \subseteq \mathscr{B}$ $(n = 1, 2, \ldots)$ *be an increasing sequence of σ-algebras, and let* $\mathscr{C}$ *be the σ-algebra generated by* $\bigcup_{n=1}^{\infty} \mathscr{C}_n$. *(In this case, we write* $\mathscr{C}_n \uparrow \mathscr{C}$.*) Then*

$$\lim_{n\to\infty} \mathbb{E}_{\mathscr{C}_n} g = \mathbb{E}_{\mathscr{C}} g \qquad \text{a.e.} \tag{27}$$

for each $g \in L_\infty(X, \mathscr{B}, \mu)$. *If* $\mathscr{A} \subseteq \mathscr{B}$ *is a finite algebra, then*

$$\lim_{n\to\infty} I(\mathscr{A}|\mathscr{C}_n) = I(\mathscr{A}|\mathscr{C}) \qquad \text{a.e.} \tag{28}$$

and

$$\lim_{n\to\infty} H(\mathscr{A}|\mathscr{C}_n) = H(\mathscr{A}|\mathscr{C}). \tag{29}$$

Proof (28) follows from (27) with $g = \chi_A$ for each $A \in \mathscr{A}$. (29) follows from (28) and the dominated convergence theorem, since

$$H(\mathscr{A}|\mathscr{C}_n) \le H(\mathscr{A}) \qquad (n = 1, 2, \ldots).$$

In order to establish (27), it is sufficient to prove

$$\lim_{n\to\infty} \mathbb{E}_{\mathscr{C}_n}\chi_A = \chi_A \qquad \text{a.e.} \tag{30}$$

for each $A \in \mathscr{C}$. For suppose that this has been established, and let f be a bounded $\mathscr{C}$-measurable function. Choose $\mathscr{C}$-measurable simple functions f_n such that

$$|f_n(x) - f(x)| \le \frac{1}{n} \qquad \text{a.e.}$$

Then, since $\mathbb{E}_{\mathscr{C}_n}$ is doubly stochastic,

$$\begin{aligned} |\mathbb{E}_{\mathscr{C}_n} f - f| &\le |\mathbb{E}_{\mathscr{C}_n} f - \mathbb{E}_{\mathscr{C}_n} f_k| + |\mathbb{E}_{\mathscr{C}_n} f_k - f_k| + |f_k - f| \\ &\le \mathbb{E}_{\mathscr{C}_n}|f - f_k| + |\mathbb{E}_{\mathscr{C}_n} f_k - f_k| + |f_k - f| \\ &\le 2/k + |\mathbb{E}_{\mathscr{C}_n} f_k - f_k|. \end{aligned}$$

According to (30), the last term tends to 0 as $n \to \infty$. Thus

$$\lim_{n\to\infty} \mathbb{E}_{\mathscr{C}_n} f = f \qquad \text{a.e.} \tag{31}$$

Finally, if $g \in L_\infty(X, \mathscr{B}, \mu)$, and if $f = \mathbb{E}_{\mathscr{C}} g$, then f is bounded and $\mathscr{C}$-measurable. From Proposition 4.2(iii),

$$\mathbb{E}_{\mathscr{C}_n} f = \mathbb{E}_{\mathscr{C}_n}\mathbb{E}_{\mathscr{C}} g = \mathbb{E}_{\mathscr{C}_n} g.$$

Thus (31) implies (27).

Now let us prove (30). Let ε and δ be arbitrary real numbers subject to $\delta > 0$, $0 < \varepsilon < 1$, and let $A \in \mathscr{C}$. Since $\mathscr{C}_n \uparrow \mathscr{C}$, there exists an integer k and a set $B \in \mathscr{C}_k$ such that

$$\mu(A \,\Delta\, B) = \mu(A \sim B) + \mu(B \sim A) < \varepsilon\delta/2. \tag{32}$$

We define successively

$$D_n = \{x : 1 - \mathbb{E}_{\mathscr{C}_n}\chi_A(x) \geq \varepsilon\} \qquad (n = 1, 2, \ldots)$$

$$F_1 = D_1 \cap B$$

$$F_{n+1} = (D_{n+1} \cap B) \sim \left(\bigcup_{j=1}^{n} F_j\right) \qquad (n = 1, 2, \ldots).$$

It follows that $D_n \in \mathscr{C}_n$ for each n, and hence (by induction) that $F_n \in \mathscr{C}_n \vee \mathscr{C}_k = \mathscr{C}_{n \vee k}$ $(n = 1, 2, \ldots)$. Here, $n \vee k$ denotes the larger of the integers n and k. Finally, letting $F = \bigcup_{n=k}^{\infty} F_n$, we have

$$F \sim A \subseteq B \sim A,$$

so that

$$\mu(B \sim A) \geq \mu(F \sim A) = \sum_{n=k}^{\infty} \mu(F_n \sim A) = \sum_{n=k}^{\infty} \int_{F_n} [1 - \chi_A]\, d\mu$$

$$= \sum_{n=k}^{\infty} \int_{F_n} [1 - \mathbb{E}_{\mathscr{C}_n}\chi_A]\, d\mu \geq \varepsilon \sum_{n=k}^{\infty} \mu(F_n) = \varepsilon\mu(F). \tag{33}$$

From (32) and (33) follows

$$\mu(F) < \delta/2. \tag{34}$$

Since

$$F = \bigcup_{n=k}^{\infty} F_n \subseteq \bigcup_{n=k}^{\infty} (B \cap D_n),$$

we now have

$$1 - \mathbb{E}_{\mathscr{C}_n}\chi_A(x) < \varepsilon \qquad (n \leq k, x \in B \sim F)$$

and

$$\mu(A \sim (B \sim F)) \leq \mu(A \sim B) + \mu(A \cap F) < \varepsilon\delta/2 + \delta/2 < \delta$$

from (32) and (34). It follows that

$$\overline{\lim_{n \to \infty}} |1 - \mathbb{E}_{\mathscr{C}_n}\chi_A(x)| \leq \varepsilon$$

for all $x \in B \sim F$ and hence, since $\delta > 0$ was arbitrary, for almost all $x \in A$. Therefore,

$$\lim_{n \to \infty} \mathbb{E}_{\mathscr{C}_n}\chi_A(x) = 1 \tag{35}$$

for almost all $x \in A$.

Since (35) holds for all $A \in \mathscr{C}$, we have

$$\lim_{n\to\infty} \mathbb{E}_{\mathscr{C}_n} \chi_{\bar{A}}(x) = \lim_{n\to\infty} [1 - \mathbb{E}_{\mathscr{C}_n} \chi_A(x)] = 1$$

or

$$\lim_{n\to\infty} \mathbb{E}_{\mathscr{C}_n} \chi_A(x) = 0 \tag{36}$$

for almost all $x \notin A$. Combining (35) and (36) yields (30). ∎

In the future we shall denote by $\bigvee_{i \in I} \mathscr{C}_i$ the smallest σ-algebra containing each of the σ-algebras $\mathscr{C}_i$ $(i \in I)$. Thus $\bigvee_{i=1}^{n} \mathscr{C}_i \uparrow \bigvee_{i=1}^{\infty} \mathscr{C}_i$, and we deduce from Theorem 4.1 and Lemma 4.1 the following corollary.

Proposition 4.8 *For each finite algebra* $\mathscr{A} \subseteq \mathscr{B}$,

$$h(\phi, \mathscr{A}) = H\left(\mathscr{A} \,\middle|\, \bigvee_{n=1}^{\infty} \phi^{-n}\mathscr{A}\right). \tag{37}$$

We could, of course, have taken (37) as the definition of $h(\phi, \mathscr{A})$. However, we shall have occasion to use both (37) and (7) in the sequel. For example, the latter is used in obtaining the following generalization of inequality (26).

Proposition 4.9 *Let* $\mathscr{A}_1, \mathscr{A}_2 \subseteq \mathscr{B}$ *be finite algebras. Then*

$$h(\phi, \mathscr{A}_1) \le h(\phi, \mathscr{A}_2) + H(\mathscr{A}_1 | \mathscr{A}_2). \tag{38}$$

Proof Exercise 8.

Theorem 4.3 (*Sinaĭ*) *If* $\mathscr{A}$ *is a finite subalgebra of* $\mathscr{B}$ *such that*

(i) $\bigvee_{n=0}^{\infty} \phi^{-n}\mathscr{A} = \mathscr{B}$, *or*
(ii) Φ *is invertible and* $\bigvee_{n=-\infty}^{\infty} \phi^{-n}\mathscr{A} = \mathscr{B}$,

then $h(\Phi) = h(\phi, \mathscr{A})$.

Proof We shall prove only (i). The second part can be proved in the same way or deduced from (i) and Theorem 4.4 below. Replacing $\mathscr{A}_2$ by $\bigvee_{j=0}^{n} \phi^{-j}\mathscr{A}$ in (38) gives

$$h(\phi, \mathscr{A}_1) \le h\left(\phi, \bigvee_{j=0}^{n} \phi^{-j}\mathscr{A}\right) + H\left(\mathscr{A}_1 \,\middle|\, \bigvee_{j=0}^{n} \phi^{-j}\mathscr{A}\right). \tag{39}$$

Since $\bigvee_{j=0}^{n} \phi^{-j}\mathscr{A} \uparrow \bigvee_{j=0}^{\infty} \phi^{-j}\mathscr{A} = \mathscr{B}$, it follows from Lemma 4.1 that the last term in (39) tends to $H(\mathscr{A}_1|\mathscr{B}) = 0$ as $n \to \infty$. On the other hand, it is easily shown (Exercise 9) that

$$h\left(\phi, \bigvee_{j=0}^{n} \phi^{-j}\mathscr{A}\right) = h(\phi, \mathscr{A})$$

for each n. Thus $h(\phi, \mathscr{A}_1) \leq h(\phi, \mathscr{A})$ for each finite algebra $\mathscr{A}_1 \subseteq \mathscr{B}$. That is

$$h(\phi, \mathscr{A}) = \sup_{\mathscr{A}_1} h(\phi, \mathscr{A}_1),$$

as required. ∎

Now suppose that $\Phi = (X, \mathscr{B}, \mu, \phi)$ is any abstract dynamical system and $\mathscr{A} \subseteq \mathscr{B}$ is any finite algebra. Then $\mathscr{B}_1 = \bigvee_{n=0}^{\infty} \phi^{-n}\mathscr{A}$ is a ϕ-invariant σ-algebra (see Section I.4). Indeed, $\phi^{-1}(\mathscr{B}_1) = \bigvee_{n=1}^{\infty} \phi^{-n}\mathscr{A} \subseteq \mathscr{B}_1$. Thus the system $\Phi_{\mathscr{A}} = (X, \mathscr{B}_1, \mu, \phi)$ is a factor of Φ. Sinaĭ's theorem says that

$$h(\phi, \mathscr{A}) = h(\Phi_{\mathscr{A}}) \tag{40}$$

for all finite algebras $\mathscr{A} \subseteq \mathscr{B}$.

We say that $\mathscr{A}$ is a *strong generator* for Φ if $\bigvee_{n=0}^{\infty} \phi^{-n}\mathscr{A} = \mathscr{B}$, and that $\mathscr{A}$ is a *generator* for Φ if $\bigvee_{n=-\infty}^{\infty} \phi^{-n}\mathscr{A} = \mathscr{B}$. We shall see that the notion of generator is very useful for evaluating $h(\Phi)$. On the other hand, we have the following consequence of Theorem 4.3.

Corollary 4.3.1 *If Φ is invertible and has a strong generator, then* $h(\Phi) = 0$.

Proof Let $\mathscr{A}$ be such a strong generator. Then

$$h(\Phi) = h(\phi, \mathscr{A}) = H\left(\mathscr{A} \,\middle|\, \bigvee_{n=1}^{\infty} \phi^{-n}\mathscr{A}\right) = H\left(\phi\mathscr{A} \,\middle|\, \bigvee_{n=0}^{\infty} \phi^{-n}\mathscr{A}\right) = 0$$

from the integrated form of (15). ∎

The next theorem, which asserts a type of continuity for the function h, may be thought of as an extension of Sinaĭ's theorem. If $\mathscr{A}$ is a generator for the invertible dynamical system Φ, and if $\mathscr{B}_1 = \bigvee_{n=0}^{\infty} \phi^{-n}\mathscr{A}$, then $\phi^m\mathscr{B}_1 \uparrow \mathscr{B}$ as $m \to \infty$. It follows that

$$\Phi = \operatorname{inv} \lim_{m \to \infty} \Phi_{\phi^m\mathscr{A}}.$$

On the other hand,

$$h(\Phi_{\phi^m\mathscr{A}}) = h(\phi, \phi^m\mathscr{A}) = h(\phi, \mathscr{A})$$

for each m. Thus Theorem 4.4 includes the second part of Theorem 4.3. It also includes various extensions of Sinaĭ's theorem given in Billingsley [7]. This was first noted by the present author in [12].

Theorem 4.4 *If* $\Phi = \text{inv}\lim_{\alpha \in J} \Phi_\alpha$, *then*

$$h(\Phi) = \lim_{\alpha} h(\Phi_\alpha). \tag{41}$$

For the proof we shall need the following lemma, which has some independent interest. The subalgebra $\mathscr{B}_0 \subseteq \mathscr{B}$ is said to be *dense* if $\mathscr{B}$ is the smallest σ-algebra containing $\mathscr{B}_0$. In this case, each $B \in \mathscr{B}$ can be approximated in $\mathscr{B}_0$, in the sense that to each $\varepsilon > 0$ corresponds a $B_\varepsilon \in \mathscr{B}_0$ with $\mu(B \,\Delta\, B_\varepsilon) < \varepsilon$.

Lemma 4.2 *Let* $\mathscr{B}_0$ *be a dense subalgebra of* $\mathscr{B}$. *Then*

$$h(\Phi) = \sup_{\mathscr{A} \subseteq \mathscr{B}_0} h(\phi, \mathscr{A}), \tag{42}$$

where the sup is taken over finite subalgebras of $\mathscr{B}_0$.

Proof Let $\mathscr{A}' \subseteq \mathscr{B}$ be any finite subalgebra. According to Proposition 4.9,

$$h(\phi, \mathscr{A}') \le h(\phi, \mathscr{A}) + H(\mathscr{A}' \,|\, \mathscr{A}),$$

where $\mathscr{A} \subseteq \mathscr{B}_0$ is any other finite algebra. We need only show that for a given $\varepsilon > 0$ there is such an $\mathscr{A}$ so that

$$H(\mathscr{A}' \,|\, \mathscr{A}) < \varepsilon. \tag{43}$$

For then it will follow that

$$h(\phi, \mathscr{A}') \le \sup_{\mathscr{A} \subseteq \mathscr{B}_0} h(\phi, \mathscr{A})$$

for each $\mathscr{A}'$, and hence

$$h(\Phi) \le \sup_{\mathscr{A} \subseteq \mathscr{B}_0} h(\phi, \mathscr{A}).$$

Let us look at the quantity

$$H(\mathscr{A}' \,|\, \mathscr{A}) = \int_X I(\mathscr{A}' \,|\, \mathscr{A}) \, d\mu.$$

For the finite algebra $\mathscr{A}$ we have (Exercise 3) that

$$\mathbb{E}_{\mathscr{A}} \chi_B = \sum_{A \in \mathscr{A}} \frac{\mu(A \cap B)}{\mu(A)} \chi_A$$

for any $B \in \mathscr{B}$. Hence

$$I(\mathscr{A}'|\mathscr{A}) = -\sum_{A' \in \mathscr{A}'} \chi_{A'} \log \mathbb{E}_{\mathscr{A}} \chi_{A'} = -\sum_{A' \in \mathscr{A}'} \chi_{A'} \log \sum_{A \in \mathscr{A}} \frac{\mu(A \cap A')}{\mu(A)} \chi_A$$

$$= -\sum_{A' \in \mathscr{A}'} \sum_{A \in \mathscr{A}} \chi_{A'} \chi_A \log \frac{\mu(A \cap A')}{\mu(A)},$$

so that

$$H(\mathscr{A}'|\mathscr{A}) = -\sum_{A' \in \mathscr{A}'} \sum_{A \in \mathscr{A}} \mu(A \cap A') \log \mu(A \cap A') + \sum_{A \in \mathscr{A}} \mu(A) \log \mu(A)$$

$$= \sum_{A \in \mathscr{A}} [F(\mu(A)) - \sum_{A' \in \mathscr{A}'} F(\mu(A \cap A'))], \tag{44}$$

where F is the continuous function defined by (5). Since $\mathscr{B}_0$ is dense in $\mathscr{B}$, we can choose a finite number of sets $A' \in \mathscr{B}_0$, one for each atom $A \in \mathscr{A}$, such that (i) they are pairwise disjoint, and (ii) for each $A \in \mathscr{A}$ there is exactly one A' with $\mu(A \,\Delta\, A') < \delta = \delta(\varepsilon/n^2)$, where $\delta(\varepsilon)$ is a modulus of continuity for F and n is the number of atoms in $\mathscr{A}$. Let $\mathscr{A}'$ be the algebra whose atoms are these sets A'. Then expression (44) contains n^2 terms each smaller in absolute value than ε/n^2, and (43) is established. ▮

Proof of Theorem 4.4 We may assume without loss of generality that $\Phi = (X, \mathscr{B}, \mu, \phi)$ and $\Phi_\alpha = (X, \mathscr{B}_\alpha, \mu, \phi)$ for each $\alpha \in J$. According to Theorem 4.2, $h(\Phi_\alpha)$ $(\alpha \in J)$ is a monotone net, and so the limit $\lim_\alpha h(\Phi_\alpha) \le +\infty$ exists and is equal to $\sup_\alpha h(\Phi_\alpha)$. Moreover, $h(\Phi_\alpha) \le h(\Phi)$ for each $\alpha \in J$, and so

$$h(\Phi) \ge \lim_\alpha h(\Phi_\alpha).$$

It only remains to prove the reverse inequality.

Let $\mathscr{B}_0 = \bigcup_{\alpha \in J} \mathscr{B}_\alpha$. Then $\mathscr{B}_0$ satisfies the hypotheses of the lemma. Let $\mathscr{A}_0 \subseteq \mathscr{B}_0$ be any *finite* subalgebra. Then there exists an $\alpha \in J$ (J is a directed set) such that $\mathscr{A}_0 \subseteq \mathscr{B}_\alpha$. But then

$$h(\phi, \mathscr{A}_0) \le \sup_{\mathscr{A} \subseteq \mathscr{B}_\alpha} h(\phi, \mathscr{A}) = h(\Phi_\alpha).$$

It follows from Lemma 4.2 that

$$h(\Phi) = \sup_{\mathscr{A} \subseteq \mathscr{B}_0} h(\phi, \mathscr{A}) \le \sup_{\alpha \in J} h(\Phi_\alpha),$$

and the proof is complete. ▮

Corollary 4.4.1 *If $\hat{\Phi}$ is the natural extension of Φ, then $h(\hat{\Phi}) = h(\Phi)$.*

Corollary 4.4.2 *If* $\Phi = \bigotimes_{\alpha \in J} \Phi_\alpha$, *then*

$$h(\Phi) = \sum_{\alpha \in J} h(\Phi_\alpha). \tag{45}$$

Proof The sum, of course, is $+\infty$ if more than countably many of the numbers $h(\Phi_\alpha)$ are positive.

It will suffice to show that

$$h(\Phi_1 \otimes \Phi_2) = h(\Phi_1) + h(\Phi_2). \tag{46}$$

If we write $\Phi = (X, \mathscr{B}, \mu, \phi)$ and $\Phi_i = (X, \mathscr{B}_i, \mu, \phi)$, then $\Phi = \Phi_1 \otimes \Phi_2$ iff $\mathscr{B} = \mathscr{B}_1 \vee \mathscr{B}_2$ and the σ-algebras $\mathscr{B}_1$ and $\mathscr{B}_2$ are *independent*, in the sense that $\mu(B_1 \cap B_2) = \mu(B_1)\,\mu(B_2)$ for all $B_1 \in \mathscr{B}_1$, $B_2 \in \mathscr{B}_2$. According to Lemma 4.2,

$$h(\Phi) = \sup_{\mathscr{A}_1, \mathscr{A}_2} h(\phi, \mathscr{A}_1 \vee \mathscr{A}_2), \tag{47}$$

where the sup is taken over all finite algebras $\mathscr{A}_1 \subseteq \mathscr{B}_1$ and $\mathscr{A}_2 \subseteq \mathscr{B}_2$. If $\mathscr{A}_1$ and $\mathscr{A}_2$ are such algebras, they are clearly independent. Moreover, $\phi^{-n}\mathscr{A}_1$ and $\phi^{-n}\mathscr{A}_2$ are independent for each n. It follows easily (Exercise 11) from this independence that

$$H(\phi^{-n}(\mathscr{A}_1 \vee \mathscr{A}_2)) = H(\phi^{-n}\mathscr{A}_1 \vee \phi^{-n}\mathscr{A}_2) = H(\phi^{-n}\mathscr{A}_1) + H(\phi^{-n}\mathscr{A}_2).$$

Dividing by n and passing to the limit gives

$$h(\phi, \mathscr{A}_1 \vee \mathscr{A}_2) = h(\phi, \mathscr{A}_1) + h(\phi, \mathscr{A}_2),$$

so that (46) follows from (47). ▌

We conclude this section by evaluating the entropy of the Bernoulli shift on k points (Example 3, Chapter I). According to Corollary 4.4.1, the two-sided and one-sided shifts have the same entropy. Let us calculate it for the two-sided shift. Thus

$$(X, \mathscr{B}, \mu) = \left(\mathop{\times}_{n=-\infty}^{\infty} X_n, \mathop{\times}_{n=-\infty}^{\infty} \mathscr{B}_n, \mathop{\times}_{n=-\infty}^{\infty} \mu_n \right),$$

where

$$X_n = \{0, 1, \ldots, k-1\}, \qquad \mathscr{B}_n = \{\text{all subsets of } X_n\},$$

$$\mu_n = \{p_0, p_1, \ldots, p_{k-1}\},$$

and $\phi(x) = y$, where $y_n = x_{n+1}$. Let $\mathscr{A}$ be the class of all cylinder sets of the form

$$A = \{x : x_0 \in A_0\},$$

where $A_0 \subseteq X_0$. Then $\mathscr{A}$ is a generator for Φ, and $\phi^{-n}\mathscr{A}$ is independent of $\phi^{-m}\mathscr{A}$ for all $n \neq m$ (since μ is the product measure). Hence

$$H\left(\bigvee_{k=0}^{n-1} \phi^{-k}\mathscr{A}\right) = nH(\mathscr{A})$$

and

$$h(\Phi) = \lim_{n\to\infty} \frac{1}{n} H\left(\bigvee_{k=0}^{n-1} \phi^{-k}\mathscr{A}\right) = H(\mathscr{A}).$$

Since the atoms of $\mathscr{A}$ are the sets

$$A = \{x : x_0 = j\} \qquad (j = 0, 1, \ldots, k-1),$$

it follows that

$$h(\Phi) = H(\mathscr{A}) = -\sum_{j=0}^{k-1} p_j \log p_j. \tag{48}$$

In particular, if $p_0 = p_1 = \cdots = p_{k-1} = 1/k$,

$$h(\Phi) = \sum_{j=0}^{k-1} \frac{1}{k} \log k = \log k.$$

This proves that, in particular, "symmetric" shifts on k points for different k are nonisomorphic.

4. TOPOLOGICAL ENTROPY

In Section II.4 we have introduced a notion of isomorphism for two classical dynamical systems $\Sigma_1 = (X_1, \sigma_1)$ and $\Sigma_2 = (X_2, \sigma_2)$. Let us say that the systems are *t-isomorphic* and write $\Sigma_1 \stackrel{t}{=} \Sigma_2$ if there exists a continuous invertible map ψ of X_1 onto X_2 such that $\psi\sigma_1 = \sigma_2\psi$.

In 1965, Adler *et al.* [4] introduced an analog $h_t(\Sigma)$ of the entropy $h(\Phi)$ for classical dynamical systems. They conjectured, and it was later proved by Goodwyn [26], Dinaburg [15], and Goodman [25], that the topological entropy $h_t(\Sigma)$ was equal to the sup of the numbers $h(\Sigma^\mu)$ where μ is a σ-invariant Borel measure on X and $\Sigma^\mu = (X, \mathscr{B}, \mu, \sigma)$. We shall take this as our definition.

Definition 4.3 Let $\Sigma = (X, \sigma)$ be a classical dynamical system and let $M(\Sigma)$ denote the class of σ-invariant normalized measures on X. The *topological entropy* $h_t(\Sigma)$ is

$$h_t(\Sigma) = \sup_{\mu \in M(\Sigma)} h(\Sigma^\mu). \tag{49}$$

A number of useful properties of topological entropy follow directly from the definition. Recall that Σ_1 is a factor of Σ_2 ($\Sigma_1 | \Sigma_2$) if there is a continuous epimorphism $\psi: X_2 \to X_1$ with $\psi\sigma_1 = \sigma_2\psi$.

Theorem 4.5 *Suppose* $\Sigma_1 | \Sigma_2$. *Then* $h_t(\Sigma_1) \le h_t(\Sigma_2)$. *In particular,* $h_t(\Sigma)$ *is a t-isomorphism invariant.*

Proof Suppose $\psi: X_2 \to X_1$ is a continuous epimorphism. Let $T_\psi : C(X_1) \to C(X_2)$ be the induced linear operator, and let $T_\psi^*: M(X_2) \to M(X_1)$ be the adjoint of T_ψ (see Section II.1). Since ψ is epic, T_ψ is monic and T_ψ^* is epic. If $\psi\sigma_2 = \sigma_1\psi$, then $T_\psi^* T_{\sigma_2}^* = T_{\sigma_1}^* T_\psi^*$, so that T_ψ^* maps the subset $M(\Sigma_2) \subseteq M(X_2)$ onto $M(\Sigma_1)$. (See Exercise II.15.) Moreover, for each $\mu \in M(\Sigma_2)$, ψ is a homomorphism of $\Sigma_2^\mu = (X_2, \mathscr{B}_2, \mu, \sigma_2)$ onto $\Sigma_1^{T_\psi^*\mu} = (X_1, \mathscr{B}_1, T_\psi^*\mu, \sigma_1)$, so that, by Theorem 4.2,

$$h(\Sigma_1^{T_\psi^*\mu}) \le h(\Sigma_2^\mu).$$

It follows that

$$h_t(\Sigma_1) = \sup_{\nu \in M(\Sigma_1)} h(\Sigma_1^\nu) = \sup_{\mu \in M(\Sigma_2)} h(\Sigma_1^{T_\psi^*\mu}) \le \sup_{\mu \in M(\Sigma_2)} h(\Sigma_2^\mu) = h_t(\Sigma_2). \quad \blacksquare$$

Theorem 4.6 *If* $\Sigma = \Sigma_1 \otimes \Sigma_2$ *is the direct product, then* $h_t(\Sigma) = h_t(\Sigma_1) + h_t(\Sigma_2)$.

Proof Suppose that $\mu \in M(\Sigma)$ is arbitrary. Let us denote by $h(\sigma, \mathscr{A}, \mu)$ the quantity $h(\sigma, \mathscr{A})$ calculated according to (7) for the abstract system Σ^μ. It follows from Eq. (16) of Section 2 and Proposition 4.6 that

$$h(\sigma, \mathscr{A}_1 \vee \mathscr{A}_2, \mu) \le h(\sigma, \mathscr{A}_1, \mu) + h(\sigma, \mathscr{A}_2, \mu) \tag{50}$$

for any finite algebras $\mathscr{A}_1, \mathscr{A}_2 \subseteq \mathscr{B}$. Since $\mathscr{B} = \mathscr{B}_1 \times \mathscr{B}_2$ is the product σ-algebra, there are canonically determined copies $\overline{\mathscr{B}}_1$ of $\mathscr{B}_1$ and $\overline{\mathscr{B}}_2$ of $\mathscr{B}_2$ in $\mathscr{B}$. (For example, $A \times X_2 \in \overline{\mathscr{B}}_1$ for each $A \in \mathscr{B}_1$.) In fact, $\mathscr{B} = \overline{\mathscr{B}}_1 \vee \overline{\mathscr{B}}_2$. Selecting $\mathscr{A}_1 \subseteq \overline{\mathscr{B}}_1$ and $\mathscr{A}_2 \subseteq \overline{\mathscr{B}}_2$ in (50) and taking the sup gives by Lemma 4.2

$$h(\Sigma^\mu) \le h(\Sigma_1^{\mu_1}) + h(\Sigma_2^{\mu_2}), \tag{51}$$

where μ_1 and μ_2 are the "marginal measures,"

$$\mu_1(A) = \mu(A \times X_2), \qquad \mu_2(B) = \mu(X_1 \times B).$$

Now if μ is the direct product $\mu_1 \times \mu_2$, then the σ-algebras $\bar{\mathscr{B}}_1$ and $\bar{\mathscr{B}}_2$ are *independent*, in the sense that

$$\mu((A \times X_2) \cap (X_1 \times B)) = \mu(A \times B)$$
$$= \mu_1(A)\, \mu_2(B) = \mu(A \times X_2) \cdot \mu(X_1 \times B)$$

for each $A \in \mathscr{B}_1$, $B \in \mathscr{B}_2$. Moreover, since $\sigma = \sigma_1 \times \sigma_2$, we have $\mu \in M(\Sigma)$ for each $\mu_1 \in M(\Sigma_1)$, $\mu_2 \in M(\Sigma_2)$. It is easily seen (Exercise 11) that we then have (50) replaced by

$$h(\sigma, \mathscr{A}_1 \vee \mathscr{A}_2, \mu) = h(\sigma, \mathscr{A}_1, \mu) + h(\sigma, \mathscr{A}_2, \mu),$$

and so

$$h(\Sigma^{\mu_1 \times \mu_2}) = h(\Sigma^{\mu_1}) + h(\Sigma^{\mu_2}). \tag{52}$$

Summing up we have for each $\mu \in M(\Sigma)$

$$h(\Sigma^{\mu}) \leq h(\Sigma_1^{\mu_1}) + h(\Sigma_2^{\mu_2}) = h(\Sigma^{\mu_1 \times \mu_2}).$$

Taking the sup first on $\mu_1 \in M(\Sigma_1)$, $\mu_2 \in M(\Sigma_2)$ and then $\mu \in M(\Sigma)$ gives

$$h_t(\Sigma) \leq h_t(\Sigma_1) + h_t(\Sigma_2) = \sup_{\mu_1, \mu_2} h(\Sigma^{\mu_1 \times \mu_2}) \leq h_t(\Sigma).$$

This completes the proof. ▮

Theorem 4.7 *If* $\Sigma = \text{inv}\lim_{\alpha \in J} \Sigma_\alpha$, *then* $h_t(\Sigma) = \lim_\alpha h_t(\Sigma_\alpha)$.

Proof For each $\alpha \in J$, Σ_α is a factor of Σ. Moreover, $\Sigma_\alpha | \Sigma_\beta$ for $\alpha \prec \beta$. Thus the net $h_t(\Sigma_\alpha)$ of real numbers is monotone nondecreasing. Hence the limit exists and

$$\lim_\alpha h_t(\Sigma_\alpha) = \sup_\alpha h_t(\Sigma_\alpha) \leq h_t(\Sigma).$$

To prove the reverse inequality, let $\mu \in M(\Sigma)$ be arbitrary. Then, as in the proof of Theorem 4.5, μ determines a measure $\mu_\alpha \in M(\Sigma_\alpha)$ for each $\alpha \in J$ such that

$$\Sigma^\mu = \text{inv}\lim_{\alpha \in J} \Sigma_\alpha^{\mu_\alpha}.$$

According to Theorem 4.4,

$$h(\Sigma^\mu) = \lim_\alpha h(\Sigma_\alpha^{\mu_\alpha}) \leq \lim_\alpha h_t(\Sigma_\alpha),$$

and so

$$h_t(\Sigma) \leq \lim_\alpha h_t(\Sigma_\alpha). \text{ ▮}$$

Corollary 4.7.1 *If* $\Sigma = \bigotimes_{\alpha \in J} \Sigma_\alpha$, *then* $h_t(\Sigma) = \sum_\alpha h_t(\Sigma_\alpha)$.

Remark It follows easily now from Furstenberg's theorem (Section II.6) that the entropy of a minimal distal system is zero (Exercise 21). This can also be proved directly in the case of a metric space (Exercise 22).

For classical dynamical systems we have the additional concepts of subsystems and sums. Let us look now at the corresponding entropy relations. The first result again follows directly from Definition 4.3.

Theorem 4.8 *If* Σ_1 *is a subsystem of* Σ_2, *then* $h_t(\Sigma_1) \leq h_t(\Sigma_2)$.

Proof Recall that Σ_1 is a subsystem of Σ_2 if there exists a continuous monomorphism $\psi: X_1 \to X_2$ such that $\psi\sigma_1 = \sigma_2\psi$. Thus we may as well assume that $X_1 \subset X_2$ is a closed subset, $\mathscr{B}_1 = \{B \cap X_1 : B \in \mathscr{B}_2\}$, and σ_1 is the restriction of σ_2 to X_1. If $\mu \in M(\Sigma_1)$, we can extend it to an invariant measure $\bar{\mu}$ on X_2 by setting

$$\bar{\mu}(A) = \mu(A \cap X_1).$$

It is easily verified (Exercise 17) that

$$h(\Sigma_1{}^{\mu}) = h(\Sigma_2{}^{\bar{\mu}}). \tag{53}$$

Hence

$$h_t(\Sigma_1) = \sup_{\mu \in M(X_1)} h(\Sigma_2{}^{\bar{\mu}}) \leq h_t(\Sigma_2). \quad \blacksquare$$

Now if $\Sigma = \Sigma_1 \oplus \Sigma_2$, then each of Σ_1 and Σ_2 is a subsystem of Σ. Thus

$$h_t(\Sigma) \geq \max\{h_t(\Sigma_1), h_t(\Sigma_2)\}. \tag{54}$$

It can be shown that the inequality in (54) can be replaced by equality. However, the proof is surprisingly involved, and we shall content ourselves with proving a related result.

Definition 4.4 Let $M_e(\Sigma)$ denote the set of ergodic measures $\mu \in M(\Sigma)$. The *ergodic entropy* $h_e(\Sigma)$ is defined by

$$h_e(\Sigma) = \sup_{\mu \in M_e(\Sigma)} h(\Sigma^{\mu}). \tag{55}$$

Proposition 4.10 *If $\Sigma = \Sigma_1 \oplus \Sigma_2$ is the direct sum, then*

$$h_e(\Sigma) = \max\{h_e(\Sigma_1), h_e(\Sigma_2)\}. \tag{56}$$

Proof Since X_1 and X_2 are σ-invariant subsets of $X = X_1 \cup X_2$, any ergodic measure μ on X must be concentrated on either X_1 or X_2. It follows as in the proof of Theorem 4.8 that $h(\Sigma^\mu)$ is either $h(\Sigma_1^{\mu_1})$ or $h(\Sigma_2^{\mu_2})$. On the other hand, an ergodic measure on X_i can be extended (in just one way) to an ergodic measure on X as in the preceding proof. Equation (56) follows from these observations and (55). ▮

Remarks 1 The set $M(\Sigma)$ is a weak*-compact, convex subset of the linear space $M(X) = C(X)^*$, and $M_e(\Sigma)$ is just exactly the set of extreme points of $M(\Sigma)$ (Exercise II.8). Moreover, as we shall show next, $h(\Sigma^\mu)$ is an affine function of μ on $M(\Sigma)$. Thus it might reasonably be conjectured that $h_e(\Sigma) = h_t(\Sigma)$. This is true, for example, if the supremum in (49) is attained for some $\mu \in M(\Sigma)$ (Exercise 20).

2 The definition of ergodic entropy is motivated by the corresponding notion of ergodic capacity in information theory, where our definition of topological entropy corresponds to stationary capacity. (See, for example, Breiman [10], in which equality of the two types of capacity is shown for symbolic dynamical systems. Since such systems are group automorphisms, this case is covered by the above remark and a theorem of K. R. Berg in the next section.)

3 In [35], Jacobs gives an integral representation for $h(\Sigma^\mu)$ which would seem to yield the equality $h_e = h_t$. However, he is concerned with abstract systems $(X, \mathscr{B}, \mu, \phi)$ with μ varying rather than with a topological setting. Thus there is in general no guarantee that his ergodic measures belong to $M_e(\Sigma)$.

4 With the historical definition of h_t (see [4]) the proof of (56) for h_t is fairly straightforward. It follows, of course, from the Goodwyn–Dinaburg–Goodman theorem that it also holds for our definition. We leave this for the interested reader to pursue, but remark that Goodwyn [27] has shown (56) does not extend to infinite sums.

Proposition 4.11 *Let Σ be a classical dynamical system. Then $h(\Sigma^\mu)$ is an affine function of $\mu \in M(\Sigma)$.*

Proof We follow [10]. For each finite subalgebra $\mathscr{A}$ of $\mathscr{B}$, let us denote by $\mathscr{A}_n$ the algebra $\bigvee_{k=0}^{n-1} \sigma^{-k}\mathscr{A}$. For $\mu_1, \mu_2 \in M(\Sigma)$, $s, t > 0$, $s + t = 1$, we have

$$\begin{aligned} h(\sigma, \mathscr{A}, s\mu_1 + t\mu_2) &= -\lim_{n\to\infty} \frac{1}{n} \sum_{A \in \mathscr{A}_n} [s\mu_1(A) + t\mu_2(A)] \log[s\mu_1(A) + t\mu_2(A)] \\ &= -\lim_{n\to\infty} \frac{1}{n} \sum_{A \in \mathscr{A}_n} s\mu_1(A) \log \mu_1(A) \\ &\quad - \lim_{n\to\infty} \frac{1}{n} \sum_{A \in \mathscr{A}_n} t\mu_2(A) \log \mu_2(A) \\ &\quad - \lim_{n\to\infty} \sum_{A \in \mathscr{A}_n} s\mu_1(A) \log[s + t\mu_2(A)/\mu_1(A)] \\ &\quad - \lim_{n\to\infty} \sum_{A \in \mathscr{A}_n} t\mu_2(A) \log[t + s\mu_1(A)/\mu_2(A)]. \end{aligned} \tag{57}$$

From the elementary inequalities

$$0 \leq \log(1 + x) \leq x \qquad (x > 0),$$

we have

$$\log s \leq \log(s + tu) \leq \log s + tu/s \qquad (u > 0).$$

Setting $u = \mu_2(A)/\mu_1(A)$, multiplying by $(s/n)\,\mu_1(A)$, and adding on A gives

$$\frac{1}{n} s \log s \leq \frac{1}{n} \sum_{A \in \mathscr{A}_n} s\mu_1(A) \log[s + t\mu_2(A)/\mu_1(A)] \leq \frac{1}{n} s \log s + \frac{t}{n}.$$

It follows that the third term on the right in (57), and likewise the fourth term, are equal to zero. Thus (57) becomes

$$h(\sigma, \mathscr{A}, s\mu_1 + t\mu_2) = sh(\sigma, \mathscr{A}, \mu_1) + th(\sigma, \mathscr{A}, \mu_2).$$

Taking the supermum over $\mathscr{A}$ yields the desired result. ∎

We conclude this section by noting that yet another entropy invariant for Σ can be defined. In Section 3.6 we defined an affine system $\tilde{\Sigma}$ associated with the classical dynamical system Σ. Clearly, $\Sigma_1 \cong \Sigma_2$ implies $\tilde{\Sigma}_1 \cong \tilde{\Sigma}_2$. Thus the *affine entropy* of Σ,

$$h_a(\Sigma) = h_t(\tilde{\Sigma}), \tag{58}$$

is an isomorphism invariant. In the next section we shall see that $h_t(\tilde{\Sigma}) = h_e(\tilde{\Sigma}) = h(\tilde{\Sigma}^m)$, where m is Haar measure.

It is not clear, in general, how $h_a(\Sigma)$ compares with $h_t(\Sigma)$ and $h_e(\Sigma)$. However, according to Theorem 3.9 and Berg's theorem, they must all coincide for systems with quasidiscrete spectrum.

5. ENTROPY OF AFFINE TRANSFORMATIONS

Let $\phi(x) = \tau(x) + a$ be an affine transformation on the compact abelian group G as in Chapter III. We denote as usual the Haar measure on G by m. We show first that the various definitions of entropy give the same value $h(\Sigma)$ for $\Sigma = (G, \tau)$, and for $\Phi = (G, \phi)$ when $h(\Sigma) < \infty$.

Theorem 4.9 (*K. R. Berg*) *Let τ be a continuous epimorphism of G, and let $\phi(x) = \tau(x) + a$. Denote $\Sigma = (G, \tau)$ and $\Phi = (G, \phi)$,*

(A) *For each $\mu \in M(\Sigma)$, $h(\Sigma^{\mu}) \leq h(\Sigma^{m})$. Thus $h_{t}(\Sigma) = h_{e}(\Sigma) = h(\Sigma^{m})$. Denote their common value by $h(\Sigma)$.*

(B) *If $h(\Sigma) < \infty$ and $\mu \in M(\Phi)$, then $h(\Phi^{\mu}) \leq h(\Phi^{m})$, and so $h_{t}(\Phi) = h_{e}(\Phi) = h(\Phi^{m})$.*

In [6], Berg proved $h(\Sigma^{\mu}) \leq h(\Sigma^{m})$ by essentially the argument given here.

Proof We prove (B) first. Let $\mu \in M(\Phi)$. Define the subalgebras $\mathscr{B}_1, \mathscr{B}_2$ of $\mathscr{B} \times \mathscr{B}$ by $\mathscr{B}_j = \pi_j^{-1}(\mathscr{B})$, where $\pi_1(x, y) = x$ and $\pi_2(x, y) = x + y$. The mapping $\rho\colon G \times G \to G \times G$ defined by $\rho(x, y) = (x, x + y)$ is a homeomorphism. Thus $\mathscr{B}_1 \vee \mathscr{B}_2 = \rho^{-1}(\mathscr{B} \times \mathscr{B}) = \mathscr{B} \times \mathscr{B}$. Moreover,

$$\pi_1\colon (G \times G, \mathscr{B}_1, m \times \mu, \tau \times \phi) \to (G, \mathscr{B}, m, \tau)$$

and

$$\pi_2\colon (G \times G, \mathscr{B}_2, m \times \mu, \tau \times \phi) \to (G, \mathscr{B}, m, \phi)$$

are isomorphisms. This is obvious for π_1 and follows for π_2 from

$$(m \times \mu)\pi_2^{-1}(A) = \int_G \int_G \chi_A(x + y)\, m(dx)\, \mu(dy) = m(A)$$

and

$$\pi_2(\tau \times \phi)(x, y) = \tau(x) + \phi(y) = \phi(x + y) = \phi\pi_2(x, y).$$

Now let $\mathscr{A}_1 \subseteq \mathscr{B}_1$ and $\mathscr{A}_2 \subseteq \mathscr{B}_2$ be finite subalgebras. Then $\mathscr{A}_1 \vee \mathscr{A}_2$ is a finite subalgebra of $\mathscr{B}_1 \vee \mathscr{B}_2 = \mathscr{B} \times \mathscr{B}$, and the union of all such algebras $\mathscr{A}_1 \vee \mathscr{A}_2$ is dense in $\mathscr{B} \times \mathscr{B}$. For each choice of $\mathscr{A}_1$ and $\mathscr{A}_2$ we have by subadditivity of entropy (cf. Exercise 11) that

$$h(\tau \times \phi, \mathscr{A}_1 \vee \mathscr{A}_2, m \times \mu) \leq h(\tau \times \phi, \mathscr{A}_1, m \times \mu) + h(\tau \times \phi, \mathscr{A}_2, m \times \mu).$$

It follows from Lemma 4.2 and the above observations that

$$h(\Sigma^{m} \otimes \Phi^{\mu}) \leq h(\Sigma^{m}) + h(\Phi^{m}).$$

According to Corollary 4.4.2, the left side is equal to $h(\Sigma^m) + h(\Phi^\mu)$. If $h(\Sigma^m) < \infty$, we can cancel it to obtain $h(\Phi^\mu) \leq h(\Phi^m)$, as asserted.

Finally, setting $a = 0$ makes $\Phi = \Sigma$, so that the above inequality becomes

$$h(\Sigma^m) + h(\Sigma^\mu) \leq 2h(\Sigma^m),$$

which gives $h(\Sigma^\mu) \leq h(\Sigma^m)$ regardless of whether or not $h(\Sigma^m) < \infty$.

The equality of $h_e(\Phi)$ and $h_t(\Phi)$ follows from $h_t(\Phi) = h(\Phi^m)$, as in Exercise 20. ▌

We turn now to the task of calculating the entropy of Φ. According to the results of Section II.4, it is sufficient in a sense to look at ergodic automorphisms and transformations with quasiperiodic spectrum.

In view of Proposition 3.7, the following theorem is a slight generalization of a result of Rohlin [52] on ergodic automorphisms of compact metrizable abelian groups.

Theorem 4.10 (*Rohlin*) *If τ is an ergodic automorphism of G and $\Sigma = (G, \tau)$ is monothetic, then Σ is a Kolmogorov system. In particular, $h(\Sigma_1) > 0$ for every nontrivial factor Σ_1 of Σ.*

Proof According to Exercise 16(b) and Theorem 3.8, it is sufficient to show that Φ_e has a σ-algebra $\mathscr{B}_0$ satisfying

(i) $\phi^{-1}(\mathscr{B}_0) \subseteq \mathscr{B}_0$,
(ii) $\bigcap_{n=1}^{\infty} \phi^{-n}(\mathscr{B}_0) = \mathscr{N}$, and
(iii) $\mathscr{B}(\bigcup_{n=1}^{\infty} \phi^n \mathscr{B}_0) = \mathscr{B}$.

Recall that $\Phi_e = (H, \tilde{\tau})$, where H is a closed subgroup of the dual of $K_d{}^\omega = \mathsf{X}_{n=-\infty}^{\infty} K_d$ (the direct product or complete direct sum) and $\tilde{\tau} = \sigma^*$ is the adjoint of the shift transformation on $K_d{}^\omega$. In fact, H is the annihilator of the group $\Delta \subseteq K_d{}^\omega$, defined as in Section III.4. Thus $\hat{H} = K_d{}^\omega/\Delta$. Now let $\Gamma_0 = \{\gamma \in K_d{}^\omega : \gamma_n = 1 \ (n > 0)\}$. It is easily seen that

(i′) $\sigma(\Gamma_0) \subseteq \Gamma_0$,
(ii′) $\bigcap_{n=1}^{\infty} \sigma^n(\Gamma_0) = \{e\}$ (e being the identity on $K_d{}^\omega$),
(iii′) $\bigcup_{n=1}^{\infty} \sigma^{-n}(\Gamma_0) + \Delta$ spans $L_2(\hat{K}_d{}^\omega)$,
(iv′) $\Gamma_0 \cap \Delta = \{e\}$.

The proof is completed by letting $\mathscr{B}_0$ be the smallest σ-algebra on which all the functions on H determined by elements of Γ_0 are measurable. ▌

Theorem 4.11 (*Seethoff*) *Let $\Sigma = (G, \tau)$, where τ is a continuous epimorphism of the compact abelian metric group G. Then $h(\Sigma) = 0$ iff Σ has quasiperiodic spectrum.*

Proof Suppose that Σ has quasiperiodic spectrum. Then Σ is distal according to Theorem 3.6. From the Remark following Corollary 4.7.1 it follows that $h(\Phi) = 0$. Alternatively, $h(\Phi) = 0$ may be proved following the same steps as in the proof of Theorem 3.6. For this purpose, we need in Step II the fact that τ on G has entropy zero when the restriction τ_H of τ to $H \subseteq G$ and the factor transformation τ/H on G/H have zero entropy. In fact (Exercise 24),

$$h(\Sigma) = h(\Sigma_H) + h(\Sigma/H). \tag{59}$$

Conversely, suppose Σ does not have quasidiscrete spectrum. Then the group $H = \Delta^\perp$ of Theorem 3.7 is nontrivial. It follows from Theorem 4.10 that $h(\Sigma_H) > 0$. Since Σ_H is a subsystem of Σ, we have by Theorems 4.8 and 4.9 that

$$h(\Sigma) = h_t(\Sigma) \geq h_t(\Sigma_H) > 0. \blacksquare$$

Theorem 4.12 *Let Φ be an ergodic affine system with quasidiscrete spectrum. Then $h(\Phi) = 0$.*

Proof Either of the alternative proofs sketched above for Σ will suffice. A third possibility is presented by Theorem 3.5, where a direct calculation may be made for $h(\hat{\Phi})$. Note that metrizability of G is not needed. ▮

Corollary 4.12.1 *Let Φ be a totally ergodic (totally minimal) abstract (classical) dynamical system with quasidiscrete spectrum. Then $h(\Phi) = 0$ ($h_t(\Phi) = 0$).*

6. McMILLAN'S THEOREM AND ENTROPY OF INDUCED SYSTEMS

According to Theorem 4.1 and Proposition 4.5, we have for any dynamical system $\Phi = (X, \mathscr{B}, \mu, \phi)$ and any finite $\mathscr{A} \subseteq \mathscr{B}$

$$h(\phi, \mathscr{A}) = H\left(\mathscr{A} \,\middle|\, \bigvee_{n=1}^{\infty} \phi^{-n}\mathscr{A}\right) = \lim_{n\to\infty} \frac{1}{n} H\left(\bigvee_{i=0}^{n-1} \phi^{-i}\mathscr{A}\right)$$

$$= \lim_{n\to\infty} \int_X \frac{1}{n} I\left(\bigvee_{i=0}^{n-1} \phi^{-i}\mathscr{A}, x\right) \mu(dx)$$

and

$$\frac{1}{n} I\left(\bigvee_{i=0}^{n-1} \phi^{-i}\mathscr{A}, x\right) = \frac{1}{n}\sum_{k=0}^{n-1} I\left(\mathscr{A} \,\middle|\, \bigvee_{i=1}^{k} \phi^{-i}\mathscr{A}, \phi^{n-k}x\right),$$

(where the empty join in the first term is taken to be the trivial algebra $\mathcal{N} = \{\phi, X\}$). According to Lemma 4.1,

$$g_k(x) = I\left(\mathscr{A} \,\middle|\, \bigvee_{i=1}^{k} \phi^{-i}\mathscr{A}, x\right) \to g(x) = I\left(\mathscr{A} \,\middle|\, \bigvee_{n=1}^{\infty} \phi^{-n}\mathscr{A}, x\right)$$

a.e. From the ergodic theorem,

$$\frac{1}{n}\sum_{k=1}^{n-1} g(\phi^{n-k}x) = \frac{1}{n}\sum_{k=1}^{n} g(\phi^k x)$$

converges a.e. to a constant, if Φ is ergodic. In fact, this constant is $\int_X g\, d\mu = h(\phi, \mathscr{A})$. We shall show that

$$\lim_{n\to\infty} \frac{1}{n}\sum_{k=0}^{n-1} g_k(\phi^{n-k}x) = h(\phi, \mathscr{A}) \tag{60}$$

in $L_1(X, \mathscr{B}, \mu)$. This result is due to McMillan [43] and proves to be essential in the calculation of the entropy of induced systems in the sense of Section I.6.

It should be noted that it has been shown by Breiman [9] that (60) holds a.e. as well. We will not need this improvement of McMillan's theorem.

Following Halmos [33], we begin by showing that the convergence $I(\mathscr{A} | \mathscr{C}_n) \to I(\mathscr{A} | \mathscr{C})$ in Lemma 4.1 holds in the norm of L_1. This will follow by a standard argument on uniform integrability from the following lemma. Recall that a sequence $\{f_n\}$ of measurable functions is *uniformly integrable* if

$$\lim_{t\to\infty} \int_{\{x: |f_n(x)| \geq t\}} |f_n(x)|\, \mu(dx) = 0$$

uniformly in n.

Lemma 4.3 *The sequence $I(\mathscr{A} | \mathscr{C}_n)$ is uniformly integrable for each choice of $\mathscr{A}$ and $\mathscr{C}_n$ $(n = 1, 2, \ldots)$.*

Proof Let $\mathscr{A} = \{A_1, \ldots, A_N\}$, and let r, s be real numbers with $0 < r \leq s$. Set

$$D_n = \{x \in X : r \leq I(\mathscr{A} | \mathscr{C}_n, x) \leq s\}$$

and

$$C_{nj} = \{x \in X : e^{-s} \leq \mathbb{E}_{\mathscr{C}_n} \chi_{A_j} \leq e^{-r}\},$$

for each $j = 1, \ldots, N$; $n = 1, 2, \ldots$. Thus $A_j \cap D_n = A_j \cap C_{nj}$, and $C_{nj} \in \mathscr{C}_n$ for each n, j. It follows that

$$\mu(A_j \cap D_n) = \mu(A_j \cap C_{nj}) = \int_{C_{nj}} \mathbb{E}_{\mathscr{C}_n} \chi_{A_j} \, d\mu \le e^{-r},$$

and so

$$\int_{A_j \cap D_n} I(\mathscr{A} | \mathscr{C}_n) \, d\mu \le s\mu(A_j \cap D_n) \le se^{-r}.$$

Summing on j gives

$$\int_{D_n} I(\mathscr{A} | \mathscr{C}_n) \, d\mu \le Nse^{-r}. \tag{61}$$

Now set $r = t + k$, $s = t + k + 1$ for each $k = 0, 1, 2, \ldots$ and any real t in (61). Adding on k gives

$$\int_{\{x : I(\mathscr{A} | \mathscr{C}_n, x) \ge t\}} I(\mathscr{A} | \mathscr{C}_n) \, d\mu \le N \sum_{k=0}^{\infty} (t + k + 1) e^{-(t+k)}. \tag{62}$$

Since the right side of (62) tends to zero as t goes to ∞, the proof is complete. ▮

Theorem 4.13 (*McMillan*) *Let $\Phi = (X, \mathscr{B}, \mu, \phi)$ be an ergodic abstract dynamical system, and let $\mathscr{A} \subseteq \mathscr{B}$ be a finite subalgebra. Then*

$$\lim_{n \to \infty} \left\| h(\phi, \mathscr{A}) - \frac{1}{n} I\left(\bigvee_{i=0}^{n-1} \phi^{-i} \mathscr{A} \right) \right\|_1 = 0.$$

Proof By uniform integrability and almost everywhere convergence, we have

$$\lim_{n \to \infty} I\left(\mathscr{A} \,\middle|\, \bigvee_{i=1}^{n} \phi^{-i} \mathscr{A} \right) = I\left(\mathscr{A} \,\middle|\, \bigvee_{n=1}^{\infty} \phi^{-n} \mathscr{A} \right) \tag{63}$$

in the norm of L_1. Again letting

$$g_k = I\left(\mathscr{A} \,\middle|\, \bigvee_{i=1}^{k} \phi^{-i} \mathscr{A} \right) \qquad \text{and} \qquad g = I\left(\mathscr{A} \,\middle|\, \bigvee_{n=1}^{\infty} \phi^{-n} \mathscr{A} \right)$$

and setting $h = h(\phi, \mathscr{A})$,

$$f_n(x) = \frac{1}{n} \sum_{k=0}^{n-1} I\left(\mathscr{A} \,\middle|\, \bigvee_{i=1}^{k} \phi^{-i} \mathscr{A}, \phi^{n-k} x \right) = \frac{1}{n} I\left(\bigvee_{i=0}^{n-1} \Phi^{-i} \mathscr{A}, x \right)$$

we have

$$\begin{aligned}\|f_n - h\|_1 &= \left\| \frac{1}{n} \sum_{k=0}^{n-1} g_k \phi^{n-k} - h \right\|_1 \\ &\leq \left\| \frac{1}{n} \sum_{k=0}^{n-1} (g_k - g)\phi^{n-k} \right\|_1 + \left\| \frac{1}{n} \sum_{k=0}^{n-1} g\phi^{n-k} - h \right\|_1 \\ &\leq \frac{1}{n} \sum_{k=0}^{n-1} \|g_k - g\|_1 + \left\| \frac{1}{n} \sum_{k=1}^{n} g\phi^k - h \right\|_1 .\end{aligned}$$

The first term tends to zero by (63) and the regularity of Cesaro sums, and the second term tends to zero by the mean ergodic theorem. ∎

Let $\mathscr{A} \subseteq \mathscr{B}$ be as above. For each $Z \in \mathscr{B}$ with $\mu(Z) > 0$ we let $Z \cap \mathscr{A}$ denote the algebra of sets $Z \cap A$ with $A \in \mathscr{A}$, and define

$$H(Z \cap \mathscr{A}) = - \sum_{A \in \mathscr{A}} \mu(Z \cap A) \log \mu(Z \cap A).$$

Let $\mathscr{A}_n$ denote the algebra $\bigvee_{i=0}^{n-1} \phi^{-i}\mathscr{A}$.

Corollary 4.13.1 *For each $\varepsilon > 0$ there exists an integer $n_0 = n_0(\varepsilon, \mathscr{A})$ such that*

$$n \geq n_0, Z \in \mathscr{A}_n \Rightarrow \left| \frac{1}{n} H(Z \cap \mathscr{A}_n) - \mu(Z)\, h(\phi, \mathscr{A}) \right| < \varepsilon. \tag{64}$$

Proof Choose n_0 such that

$$n \geq n_0 \Rightarrow \left\| h(\phi, \mathscr{A}) - \frac{1}{n} I(\mathscr{A}_n) \right\|_1 < \varepsilon.$$

For such an n and any $Z \in \mathscr{A}_n$, $Z \cap A = A$ or $\varnothing$ for each $A \in \mathscr{A}_n$, so that

$$\begin{aligned}\left| \frac{1}{n} H(Z \cap \mathscr{A}_n) - \mu(Z) h(\phi, \mathscr{A}) \right| &= \left| \int_Z \left[\frac{1}{n} I(\mathscr{A}_n, x) - h(\phi, \mathscr{A}) \right] \mu(dx) \right| \\ &\leq \int_X \left| \frac{1}{n} I(\mathscr{A}_n) - h(\phi, \mathscr{A}) \right| d\mu < \varepsilon. \end{aligned}$$ ∎

Remark The function $f(t) = -t \log t$ is monotone increasing on $[0, e^{-1}]$, so that whenever Z and $\mathscr{A}$ are such that $\mu(Z \cap A) \leq e^{-1}$ for each $A \in \mathscr{A}$, we have for each measurable set $W \subseteq Z$ that $H(W \cap \mathscr{A}) \leq H(Z \cap \mathscr{A})$.

Corollary 4.13.2 *For each* $\varepsilon > 0$ *and* $Z \in \mathscr{B}$ *with* $\mu(Z) < \min\{\varepsilon, e^{-1}\}$ *there exists an integer* $n_1 = n_1(\varepsilon, \mathscr{A}, Z)$ *such that*

$$n \geq n_1,\ W \subseteq Z \Rightarrow H(W \cap \mathscr{A}_n) < n\varepsilon. \tag{65}$$

Proof Let $\mathscr{C} = \{Z, X \sim Z\}$. From the previous corollary and subsequent remark, and the subadditivity of entropy, we have

$$\begin{aligned} n &\geq n_0(\varepsilon, \mathscr{A} \vee \mathscr{C}) \\ &\Rightarrow \frac{1}{n} H(W \cap \mathscr{A}_n) \leq \frac{1}{n} H(Z \cap \mathscr{A}_n) \leq \frac{1}{n} H(Z \cap (\mathscr{A} \vee \mathscr{C})_n) \\ &\qquad \leq \mu(Z)\, h(\phi, \mathscr{A} \vee \mathscr{C}) + \varepsilon \leq \varepsilon[h(\phi, \mathscr{A}) + h(\phi, \mathscr{C}) + 1] \\ &\qquad \leq \varepsilon[h(\phi, \mathscr{A}) + \log 2 + 1]. \end{aligned}$$

Thus we can take $n_1(\varepsilon, \mathscr{A}, Z) = n_0(\varepsilon/[h(\phi, \mathscr{A}) + \log 2 + 1], \mathscr{A} \vee \mathscr{C})$. ∎

Now we are prepared to use Corollaries 4.13.1 and 4.13.2 to evaluate the entropy $h(\Phi_Y)$ of the induced system Φ_Y as defined in Section I.6. First, however, it is convenient to make a slight extension of these results.

Up until now we have considered only finite algebras $\mathscr{A}$ in the construction of entropy. However, it is not hard to show (Exercise 26) that the same results, including McMillan's theorem and its corollaries, are valid for σ-algebras $\mathscr{A}$ generated by countable partitions of X and satisfying

$$H(\mathscr{A}) = -\sum_{A \in \mathscr{A}} \mu(A) \log \mu(A) < \infty,$$

where the sum now is an infinite series.

In the proof of Theorem 1.9 we introduced the sets

$$A_n = \{x \in Y : F(x) = n\},$$

where $F(x) = n_Y(x)$ is the smallest positive integer n such that $\phi^n(x) \in Y$. Let $\bar{\mathscr{A}}^0$ be the σ-algebra whose atoms are the A_n so defined and the single set $X \sim Y$.

Lemma 4.4 $H(\bar{\mathscr{A}}^0) < \infty$ *and*

$$\int_Y F(x)\, \mu(dx) = \sum_{k=1}^{\infty} k\mu(A_k) = 1. \tag{66}$$

Proof Since

$$H(\bar{\mathscr{A}}^0) = -\mu(X \sim Y) \log \mu(X \sim Y) - \sum_{k=1}^{\infty} \mu(A_k) \log \mu(A_k),$$

it follows from elementary considerations of positive-term series that the finiteness of $H(\overline{\mathscr{A}}^0)$ follows from the convergence of the infinite series in (66). The proof of (66) is left to the exercises. ∎

Remark The equality in (66) is usually known as Kac's theorem and is a refinement of the recurrence theorem (Theorem 1.7). It says that the "mean recurrence time" of the set Y is inversely proportional to the measure of Y,

$$\sum_{k=1}^{\infty} k \frac{\mu(A_k)}{\mu(Y)} = \frac{1}{\mu(Y)}.$$

Let us define a sequence ρ_n of integer-valued functions on Y by

$$\rho_n(x) = \sum_{k=0}^{n-1} F(\phi_Y{}^k x). \tag{67}$$

Thus $\rho_n(x)$ is the number of steps in the sequence $\phi^k x$ until the nth return to the set Y.

For each σ-algebra $\mathscr{A} \subseteq Y \cap \mathscr{B}$ of measurable subsets of Y, let us denote by $\overline{\mathscr{A}}$ the σ-algebra generated by $\mathscr{A}$ and the single set $X \sim Y$. Let $\mathscr{A}$ be such a σ-algebra generated by a countable partition of Y, and assume that $\mathscr{A}^0 = Y \cap \overline{\mathscr{A}}^0 \subseteq \mathscr{A}$. Thus each atom of $\mathscr{A}^0$ is a union of atoms of $\mathscr{A}$, and each atom of $\mathscr{A}$ is entirely contained in some atom of $\mathscr{A}^0$. Moreover, each atom of $\mathscr{A}_n = \bigvee_{k=0}^{n-1} \phi_Y^{-k}\mathscr{A}$ is contained in an atom of $\mathscr{A}_n{}^0 = \bigvee_{k=0}^{n-1} \phi_Y^{-k}\mathscr{A}^0$. Clearly, ρ_n is constant on these atoms. Thus

(α) *ρ_n is constant on the atoms of $\mathscr{A}_n$*. Let A be such an atom. It is only slightly less obvious that

(β) *$\rho_n(A) = m$, $A \in \dot{\mathscr{A}}_n \Rightarrow A$ is an atom of $\overline{\mathscr{A}}_m = \bigvee_{k=0}^{m-1} \phi^{-k}\overline{\mathscr{A}}$.*

Theorem 4.14 (*Abramov*) *Let $\Phi = (X, \mathscr{B}, \mu, \phi)$ be an ergodic abstract dynamical system, and suppose $Y \in \mathscr{B}$ with $\mu(Y) > 0$. Then $h(\Phi) = h(\Phi_Y)\,\mu(Y)$.*

Proof [1] According to Proposition 1.12, Φ_Y is ergodic. From the ergodic theorem and Lemma 4.4,

$$\lim_{n\to\infty} \frac{1}{n}\rho_n(x) = \frac{1}{\mu(Y)}\int_Y F\,d\mu = \frac{1}{\mu(Y)} \qquad \text{a.e. on } Y.$$

It follows that $(1/n)\rho_n \to 1/\mu(Y)$ almost uniformly. That is, for each $\varepsilon > 0$ there is a measurable set Y_ε with $\mu(Y \sim Y_\varepsilon) < \varepsilon$ and $(1/n)\rho_n \to 1/\mu(Y)$ uniformly on Y_ε. Let $\mathscr{A}$ be a σ-algebra of measurable subsets of Y, generated by a countable partition of Y and containing $\mathscr{A}^0$. Choose N such that $n \geq N$ implies

(i) $|(1/n)\rho_n(x) - 1/\mu(Y)| < \varepsilon$ for $x \in Y_\varepsilon$,

(ii) $|(1/n)H(B \cap \overline{\mathscr{A}}_n) - h(\phi, \overline{\mathscr{A}})\,\mu(B)| < \varepsilon$ for all $B \in \overline{\mathscr{A}}_n = \bigvee_{k=0}^{n-1} \phi^{-k}\overline{\mathscr{A}}$ (Corollary 4.13.1),

(iii) $|(1/n)H_Y(B \cap \mathscr{A}_n) - h(\phi_Y, \mathscr{A})\,\mu(B)/\mu(Y)| < \varepsilon$ for all $B \in \mathscr{A}_n = \bigvee_{k=0}^{n-1} \phi_Y^{-k}\mathscr{A}$ (Corollary 4.13.1),

(iv) $(1/n)H(B \cap \mathscr{A}_n) < \varepsilon$ for all $B \subseteq Y \sim Y_\varepsilon$ (Corollary 4.13.2), and

(v) $1/n < \varepsilon$,

where H_Y is calculated by replacing $\mu(A)$ by $\mu(A)/\mu(Y)$ in H.

For each $n \geq N$ let Y_n be the union of those atoms of $\mathscr{A}_n$ that intersect Y_ε in a set of positive measure. According to (α), inequality (i) is valid for all $x \in Y_n$. Let $m_1 = \min_{x \in Y_n} \rho_n(x)$, $m_2 = \max_{x \in Y_n} \rho_n(x)$. (Note that $m_2 < \infty$ because of (i) and the previous sentence.) According to (β), $Y_n \cap \overline{\mathscr{A}}_{m_1} \subseteq Y_n \cap \mathscr{A}_n \subseteq Y_n \cap \overline{\mathscr{A}}_{m_2}$. Thus

$$H_X(Y_n \cap \overline{\mathscr{A}}_{m_1}) \leq H_X(Y_n \cap \mathscr{A}_n) \leq H_X(Y_n \cap \overline{\mathscr{A}}_{m_2}). \tag{68}$$

Now note that

$$\begin{aligned} H_X(Y_n \cap \mathscr{A}_n) &= -\sum_{A \in \mathscr{A}_n} \mu(A \cap Y_n) \log \mu(A \cap Y_n) \\ &= \mu(Y)\, H_Y(Y_n \cap \mathscr{A}_n) + \mu(Y_n) \log \mu(Y) \end{aligned}$$

becomes

$$H_X(Y_n \cap \overline{\mathscr{A}}_{m_1}) \leq \mu(Y)\, H_Y(Y_n \cap \mathscr{A}_n) + \mu(Y_n) \log \mu(Y) \leq H_X(Y_n \cap \overline{\mathscr{A}}_{m_2}). \tag{69}$$

From (69) and (i) we have

$$\begin{aligned} \frac{1}{n} H_Y(Y_n \cap \mathscr{A}_n) &\leq \frac{m_2}{n} \left[\frac{1}{\mu(Y)} \frac{1}{m_2} H_X(Y_n \cap \overline{\mathscr{A}}_{m_2}) - \frac{1}{m_2} \frac{\mu(Y_n)}{\mu(Y)} \log \mu(Y)\right] \\ &\leq \left[\frac{1}{\mu(Y)} + \varepsilon\right] \left[\frac{1}{\mu(Y)} \frac{1}{m_2} H_X(Y_n \cap \mathscr{A}_{m_2}) \right. \\ &\qquad \left. - \frac{1}{m_2} \frac{\mu(Y_n)}{\mu(Y)} \log \mu(Y)\right]. \end{aligned} \tag{70}$$

Likewise, and by (ii),

$$\begin{aligned} \frac{1}{n} H_Y(Y_n \cap \mathscr{A}_n) &\geq \left[\frac{1}{\mu(Y)} - \varepsilon\right] \left[\frac{1}{\mu(Y)} \frac{1}{m_1} H_X(Y_n \cap \overline{\mathscr{A}}_{m_1}) \right. \\ &\qquad \left. - \frac{1}{m_1} \frac{\mu(Y_n)}{\mu(Y)} \log \mu(Y)\right] \\ &\geq \left[\frac{1}{\mu(Y)} - \varepsilon\right] \left[\frac{1}{\mu(Y)} \left(\frac{1}{m_2} H_X(Y_n \cap \overline{\mathscr{A}}_{m_2}) - 2\varepsilon\right) \right. \\ &\qquad \left. - \frac{1}{m_1} \frac{\mu(Y_n)}{\mu(Y)} \log \mu(Y)\right] \end{aligned} \tag{71}$$

Let $\mathscr{C} = \{Y, X \sim Y\}$. Then

$$\begin{aligned} H_X(Y \cap \overline{\mathscr{A}}_{m_2}) &\leq H_X(Y \cap (\overline{\mathscr{A}} \vee \mathscr{C})_{m_2}) \\ &= H_X(Y_n \cap \overline{\mathscr{A}}_{m_2}) + H_X((Y \sim Y_n) \cap \overline{\mathscr{A}}_{m_2}). \end{aligned}$$

Thus by (iv)

$$H_X(Y_n \cap \overline{\mathscr{A}}_{m_2}) \leq H_X(Y \cap \overline{\mathscr{A}}_{m_2}) \leq H_X(Y_n \cap \overline{\mathscr{A}}_{m_2}) + \varepsilon. \tag{72}$$

Noting that $m_1, m_2 \to \infty$ as $n \to \infty$, it is seen from (70)–(72) that

$$\limsup_{n\to\infty} \frac{1}{\varepsilon} \left| \frac{\mu(Y)}{n} H_Y(Y_n \cap \mathscr{A}_n) - \frac{1}{m_2\,\mu(Y)} H_X(Y_n \cap \overline{\mathscr{A}}_{m_2}) \right| < \infty. \tag{73}$$

Since we have by (73), (ii), (iii), and (iv) similar statements regarding the differences of consecutive terms of

$$h(\phi, \overline{\mathscr{A}}), \quad \frac{1}{m_2\,\mu(Y)} H_X(Y \cap \overline{\mathscr{A}}_{m_2}), \quad \frac{1}{m_2\,\mu(Y)} H_X(Y_n \cap \overline{\mathscr{A}}_{m_2}),$$

$$\frac{\mu(Y)}{n} H_Y(Y_n \cap \mathscr{A}_n), \quad \frac{\mu(Y)}{n} H_Y(Y \cap \mathscr{A}_n), \quad \mu(Y)\, h(\phi_Y, \mathscr{A}),$$

it follows that

$$|h(\phi, \overline{\mathscr{A}}) - \mu(Y)\, h(\phi_Y, \mathscr{A})| < M\varepsilon$$

for some constant M independent of ε. Since ε was arbitrary,

$$h(\phi, \overline{\mathscr{A}}) = \mu(Y)\, h(\phi_Y, \mathscr{A}). \tag{74}$$

Clearly, the union of all σ-algebras $\mathscr{A}$ of the form described is dense in $Y \cap \mathscr{B}$. Since Φ is ergodic, it follows easily that the union of the associated σ-algebras $\overline{\mathscr{A}}$ is dense in $\mathscr{B}$. Thus, by Lemma 4.2, the statement of the theorem follows from (74). ▮

EXERCISES

Isomorphism

1. (a) Linear operators T_1 and T_2 on Hilbert spaces H_1 and H_2 are *isomorphic* if there exists a linear operator S: $H_1 \to H_2$ such that (i) S is invertible, (ii) $\|Sf\| = \|f\|$ for all $f \in H_1$, and (iii) the diagram

$$\begin{array}{ccc} H_1 & \xrightarrow{T_1} & H_1 \\ \Big\downarrow S & & \Big\downarrow S \\ H_2 & \xrightarrow{T_2} & H_2 \end{array}$$

commutes. If $\Phi_1 \stackrel{\mathrm{m}}{=} \Phi_2$, show that T_{ϕ_1} and T_{ϕ_2} are isomorphic.

(b) Affine transformations ϕ_1 and ϕ_2 on compact abelian groups G_1 and G_2 are *g-isomorphic* ($\phi_1 \stackrel{\mathrm{g}}{=} \phi_2$) if there exists an affine transformation $\psi: G_1 \to G_2$ such that (i) ψ is invertible, (ii) ψ is continuous, and (iii) the diagram

$$\begin{array}{ccc} G_1 & \xrightarrow{\phi_1} & G_1 \\ \Big\downarrow \psi & & \Big\downarrow \psi \\ G_2 & \xrightarrow{\phi_2} & G_2 \end{array}$$

commutes. Show that $\Phi_1 \stackrel{\mathrm{m}}{=} \Phi_2$ implies $\tilde{\phi}_1 \stackrel{\mathrm{g}}{=} \tilde{\phi}_2$. In particular, any two choices of $\alpha \in G$ in Eq. (23) of Section III.6 yield g-isomorphic affine transformations $\tilde{\phi}$.

Conditional Expectation

2. Prove Proposition 4.1.

3. **(a)** The conditional expectation of f given $\mathscr{C}$ is usually defined in the following way. Show that the two definitions are equivalent.

Definition 4.1* For each $f \in L_1(X, \mathscr{B}, \mu)$, $\mathbb{E}_{\mathscr{C}} f = g$ iff
(i) g is $\mathscr{C}$-measurable, and
(ii) $\int_C g \, d\mu = \int_C f \, d\mu$ for all $C \in \mathscr{C}$.

(b) For a finite algebra $\mathscr{A}$ with atoms $A_1, \ldots, A_k$, show that

$$\mathbb{E}_{\mathscr{A}} f = \sum_{A \in \mathscr{A}} \frac{1}{\mu(A)} \int_A f \, d\mu \, \chi_A.$$

4. Let ψ be a measure-preserving transformation of the normalized measure space $(X_1, \mathscr{B}_1, \mu_1)$ onto the normalized measure space $(X_2, \mathscr{B}_2, \mu_2)$. The two spaces may coincide. Show that the induced linear operator T_ψ satisfies

$$T_\psi T_\psi^* = \mathbb{E}_{\psi^{-1}(\mathscr{B}_2)}.$$

Thus the theory of conditional expectations is coextensive with the theory of noninvertible measure-preserving transformations.

5. Complete the proof of Proposition 4.3.

Entropy

6. Show that F as defined by Eq. (5) satisfies the hypotheses of Proposition 4.4.

7. Complete the proof of Proposition 4.7.

8. (a) Show that $H(\mathscr{A}|\mathscr{C}) = 0$ iff $\mathscr{A} \subseteq \mathscr{C}$.

(b) For finite algebras $\mathscr{A}, \mathscr{A}' \subseteq \mathscr{B}$ and any σ-algebra $\mathscr{C} \subseteq \mathscr{B}$,

$$H(\mathscr{A} \vee \mathscr{A}'|\mathscr{C}) = H(\mathscr{A}|\mathscr{C}) + H(\mathscr{A}'|\mathscr{A} \vee \mathscr{C}) \leq H(\mathscr{A}|\mathscr{C}) + H(\mathscr{A}'|\mathscr{C}).$$

(c) For any finite algebras $\mathscr{A}_1, \mathscr{A}_2 \subseteq \mathscr{B}$

$$\begin{aligned} H\left(\bigvee_{i=0}^{n-1} \phi^{-i}\mathscr{A}_1\right) &\leq H\left[\left(\bigvee_{i=0}^{n-1} \phi^{-i}\mathscr{A}_1\right) \vee \left(\bigvee_{i=0}^{n-1} \phi^{-i}\mathscr{A}_2\right)\right] \\ &= H\left(\bigvee_{i=0}^{n-1} \phi^{-i}\mathscr{A}_2\right) + H\left(\bigvee_{i=0}^{n-1} \phi^{-i}\mathscr{A}_1 \,\middle|\, \bigvee_{j=0}^{n-1} \phi^{-j}\mathscr{A}_2\right) \\ &\leq H\left(\bigvee_{i=0}^{n-1} \phi^{-i}\mathscr{A}_2\right) + \sum_{i=0}^{n-1} H(\phi^{-i}\mathscr{A}_1|\phi^{-i}\mathscr{A}_2) \\ &= H\left(\bigvee_{i=0}^{n-1} \phi^{-i}\mathscr{A}_2\right) + nH(\mathscr{A}_1|\mathscr{A}_2). \end{aligned}$$

(d) $h(\phi, \mathscr{A}_1) \leq h(\phi, \mathscr{A}_2) + H(\mathscr{A}_1|\mathscr{A}_2)$.

9. (a) Deduce from Exercise 8(b) that

$$\mathscr{A}_2 \subseteq \mathscr{C} \Rightarrow H(\mathscr{A}_1 \vee \mathscr{A}_2|\mathscr{C}) = H(\mathscr{A}_1|\mathscr{C}).$$

(b) For any finite $\mathscr{A}_1, \mathscr{A}_2 \subseteq \mathscr{B}$ and any $\mathscr{C} \subseteq \mathscr{B}$,

$$H(\mathscr{A}_1 \vee \mathscr{A}_2|\mathscr{C} \vee \mathscr{A}_2) \leq H(\mathscr{A}_1|\mathscr{C}).$$

(c) For any finite $\mathscr{A} \subseteq \mathscr{B}$ and any k,

$$h\left(\phi, \bigvee_{j=0}^{k} \phi^{-j}\mathscr{A}\right) = H\left(\bigvee_{j=0}^{k} \phi^{-j}\mathscr{A} \,\middle|\, \bigvee_{n=1}^{\infty} \phi^{-n}\mathscr{A}\right) = H\left(\mathscr{A} \,\middle|\, \bigvee_{n=1}^{\infty} \phi^{-n}\mathscr{A}\right) = h(\phi, \mathscr{A}).$$

10. Show that $h(\phi, \mathscr{A}) = h(\phi, \phi^{-1}\mathscr{A})$ and, if ϕ is invertible, $h(\phi, \mathscr{A}) = h(\phi, \phi\mathscr{A})$.

11. Suppose that $\mathscr{A}_1, \mathscr{A}_2 \subseteq \mathscr{B}$ are *independent*, that is, $\mu(A_1 \cap A_2) = \mu(A_1)\,\mu(A_2)$ for $A_1 \in \mathscr{A}_1$, $A_2 \in \mathscr{A}_2$. Show that

$$H(\mathscr{A}_1 \vee \mathscr{A}_2) = H(\mathscr{A}_1) + H(\mathscr{A}_2).$$

Conversely, if $\mathscr{A}_1$ and $\mathscr{A}_2$ are not independent, then

$$H(\mathscr{A}_1 \vee \mathscr{A}_2) < H(\mathscr{A}_1) + H(\mathscr{A}_2).$$

Calculation of Entropy

12. (*Markov shift*) Let $X = \mathsf{X}_{n=-\infty}^{\infty} X_n$, where $X_n = \{0, 1, \ldots, k-1\}$, and define $\mathscr{B}$ and ϕ as for the Bernoulli shift. A ϕ-invariant measure μ is defined on $(X, \mathscr{B})$ as follows. We suppose given nonnegative numbers p_j $(j = 0, 1, \ldots, k-1)$ and p_{ij} $(i, j = 0, 1, \ldots, k-1)$ such that

$$\sum_k p_k = 1, \qquad \sum_k p_{ik} = 1, \qquad \sum_k p_k p_{kj} = p_j \qquad (i, j = 0, \ldots, k-1).$$

The measure μ is defined on cylinder sets by

$$\mu\{x : x_t = i_0, x_{t-1} = i_1, \ldots, x_{t-n} = i_n\} = p_{i_0} p_{i_0 i_1} \cdots p_{i_{n-1} i_n}.$$

This measure is then extended (as an inverse limit of measures on finite products) to a measure on $(X, \mathscr{B})$. Show as for the Bernoulli shift, that the algebra $\mathscr{A}$ of cylinder sets based on X_0 is a generator, and that

$$h(\Phi) = h(\phi, \mathscr{A}) = -\sum_i p_i \sum_j p_{ij} \log p_{ij}.$$

13. (*Translations have zero entropy*)

(a) For any finite $\mathscr{A} \subseteq \mathscr{B}$ and any positive integer k,

$$h\left(\phi^k, \bigvee_{j=0}^{k-1} \phi^{-j}\mathscr{A}\right) = k\, h(\phi, \mathscr{A}).$$

(b) For any $k \in Z$, let $\Phi^k = (X, \mathscr{B}, \mu, \phi^k)$. Then

$$h(\Phi^k) = |k|\, h(\Phi).$$

(c) If $X = K$ is the circle group, and if ϕ is multiplication by a root of unity, then $\phi^k = I$ for some k, and hence $h(\Phi) = 0$.

(d) If $X = K$, if ϕ is rotation through an irrational multiple of π, and if $\mathscr{A}$ is the algebra whose atoms are the upper and lower half-circles, then $\mathscr{A}$ is a generator for Φ, $\bigvee_{n=1}^{\infty} \phi^{-n}\mathscr{A} = \mathscr{B}$. Hence

$$h(\Phi) = H(\mathscr{A} \mid \mathscr{B}) = 0.$$

(e) If $X = K^n$ is the n-torus, and if $\phi(x) = x + a$, then $h(\Phi) = 0$.

14. **(a)** If X is a finite set, then $h(\Phi) = 0$.
(b) If Φ is ergodic and $h(\Phi) > 0$, then X is nonatomic.

Deterministic and Nondeterministic Systems

15. **(a)** Suppose $h(\Phi) = 0$ and $A \in \mathscr{B}$. Let $\mathscr{A} = \{\varnothing, A, \tilde{A}, X\}$. Then $h(\phi, \mathscr{A}) = 0$. Deduce from this and Exercise 8(a) that $A \in \phi^{-1}(\mathscr{B})$, and hence that Φ is invertible.

(b) If $h(\Phi) \neq 0$, there exists a finite algebra $\mathscr{A}$ such that the factor $\Phi_{\mathscr{A}} = \{X, \bigvee_{n=0}^{\infty} \phi^{-n}\mathscr{A}, \mu, \phi\}$ of Φ is noninvertible.

(c) The system Φ is said to be *deterministic* if $\mathscr{A} \subseteq \bigvee_{n=1}^{\infty} \phi^{-n}\mathscr{A}$ for each finite algebra $\mathscr{A} \subseteq \mathscr{B}$. Show that the following are equivalent:
(i) $h(\Phi) = 0$.
(ii) Φ is deterministic.
(iii) Every factor of Φ is invertible.

16. Let $\Phi = (X, \mathscr{B}, \mu, \phi)$ be a dynamical system, and let $\mathscr{B}_{\mathrm{d}}$ be the join of all ϕ-invariant σ-algebras $\mathscr{B}_0 \subseteq \mathscr{B}$ such that $\Phi_0 = (X, \mathscr{B}_0, \mu, \phi)$ is deterministic (has entropy zero). Equivalently, let Φ_{d} be the sup of the deterministic factors of Φ. If $\mathscr{B}_{\mathrm{d}} = \mathscr{N}$ is trivial, Φ is *completely nondeterministic*.

(a) Show that $\Phi_{\mathrm{d}} = (X, \mathscr{B}_{\mathrm{d}}, \mu, \phi)$ is deterministic.
(b) The following are equivalent.
(i) Φ is completely nondeterministic.
(ii) Every factor of Φ has positive entropy. (We say Φ has *completely positive entropy*.)
(iii) Every factor of Φ has a noninvertible factor.
(iv) The natural extension of Φ is a Kolmogorov system (see Exercise I.25).

Topological Entropy

17. Verify Eq. (53).

18. If $\Sigma = (X, \sigma)$ is a classical dynamical system, and if $\Sigma^k = (X, \sigma^k)$, show that $h_{\mathrm{t}}(\Sigma^k) = |k| h_{\mathrm{t}}(\Sigma)$.

19. Show directly from Definition 4.3 and Eq. (48) that the topological entropy of the k-shift is $\log k$.

20. Suppose that $h_{\mathrm{t}}(\Sigma) = h(\Sigma^{\mu})$ for some $\mu \in M(\Sigma)$.
(a) Show that the set

$$A = \{\nu \in M(\Sigma) : h(\Sigma^{\nu}) = h_{\mathrm{t}}(\Sigma)\}$$

is weak*-compact, convex, and nonempty.

(b) Show that A is an extremal set, in the sense that

$$\begin{matrix} sv_1 + tv_2 \in A \\ 0 < s, t, s + t = 1 \end{matrix} \Rightarrow v_1, v_2 \in A.$$

(c) Conclude that any extreme point of A is an extreme point of $M(\Sigma)$, and hence that there exists an *ergodic* μ with $h_t(\Sigma) = h(\Sigma^\mu)$.

21. **(a)** If Φ_1 is an algebraic extension of Φ, and if $h_t(\Phi) = 0$, show that $h_t(\Phi_1) = 0$.

(b) Show that the class $\mathscr{C}$ of dynamical systems with zero entropy satisfies the hypotheses of Furstenberg's theorem (Section II.6). Conclude that every minimal distal system has zero entropy.

22. (*W. Parry*) Let (X, d) be a compact metric space, and let σ be a distal homeomorphism of X. Let $\mu \in M(\Sigma)$ be arbitrary. For some $z \in X$ choose a sequence S_n of spheres centered at z with $\operatorname{diam}(S_n) \to 0$.

(a) If $\hat{S}_n = S_n \sim \{z\}$, show that $\hat{S}_n \downarrow \phi$. Hence $\mu(\hat{S}_n) \to 0$, and we may assume by passing to a subsequence that $\mu(\hat{S}_n) \leq r^n$ for some (arbitrary) r satisfying $0 < r < 1/e$.

(b) Let $A_0 = (S_0 \sim S_1) \cup \{z\}$ and $A_n = S_n \sim S_{n+1}$ $(n \geq 1)$. Let $\mathscr{A}_n = \{A_0, \ldots, A_{n-1}, \hat{S}_n\}$, so that $\mathscr{A}_n \uparrow \mathscr{A}$, where $\mathscr{A}$ is the σ-field generated by the countable number of atoms $A_0, A_1, \ldots, A_n, \ldots$. Let $x, y \in X$. If there exists a sequence n_k of integers such that $\sigma^{n_k}x, \sigma^{n_k}y \in \mathscr{A}_k \subseteq S_k$, show by distality that $x = y$. On the other hand, by minimality there exist for each k arbitrarily large integers n with $\sigma^n x \in S_k$. Conclude that $\bigvee_{n=1}^{\infty} \sigma^{-n}\mathscr{A} = \mathscr{B}$.

(c) Show from (b) that $h(\Sigma^\mu) = \lim_{n\to\infty} h(\sigma, \mathscr{A}_n, \mu)$.

(d) Show that

$$\begin{aligned} h(\sigma, \mathscr{A}_n, \mu) \leq H(\mathscr{A}_n) &\leq -\mu(A_0) \log \mu(A_0) - \sum_{n=1}^{\infty} nr^n \log r \\ &\leq (r-1)\log(1-r) + \sum_{n=1}^{\infty} nr^n \\ &= (r-1)\log(1-r) + r/(1-r)^2. \end{aligned}$$

(e) Conclude from (c) and (d), the definition of h_t, and the arbitrariness of r and μ, that $h_t(\Sigma) = 0$.

Affine Systems

23. Complete the details of the proof of Theorem 4.9.

24. **(a)** Verify Eq. (59).

(b) Carry out the alternative proof that $h(\Sigma) = 0$ when Σ has quasi-periodic spectrum.

25. (*Algebraic entropy*) Let τ be an epimorphism of the compact abelian group G. For any finite subgroup F of $\hat{G}$ define $H(F) = \log \operatorname{ord}(F)$, where $\operatorname{ord}(F)$ denotes the order of F.

(a) Show that H is subadditive on the lattice $\mathcal{T}$ of finite subgroups of $\hat{G}$, that is,

$$H(F_1 \vee F_2) \leq H(F_1) + H(F_2).$$

(b) Show that $H(\tau^* F) = H(F)$ for each $F \in \mathcal{T}$.

(c) Show that

$$H\left(\bigvee_{i=0}^{n+m-1} \tau^{*i}F\right) \leq H\left(\bigvee_{i=0}^{n-1} \tau^{*i}F\right) + H\left(\bigvee_{i=0}^{m-1} \tau^{*i}F\right)$$

and hence that

$$h(\tau^*, F) = \lim_{n\to\infty} \frac{1}{n} H\left(\bigvee_{i=0}^{n-1} \tau^{*i}F\right) = \inf_n \frac{1}{n} H\left(\bigvee_{i=0}^{n-1} \tau^{*i}F\right)$$

exists. We define the *algebraic entropy* of $\Sigma = (G, \tau)$ by

$$h_a(\Sigma) = \sup_{F \in \mathcal{T}} h(\tau^*, F).$$

(d) Show that $h_a(\Sigma) = h_a(\Sigma/\Sigma_0)$, where $\Sigma/\Sigma_0 = (G/G_0, \tau/G_0)$ is an algebraic factor of Σ, and G_0 is the connected component of $0 \in G$. (Hint: Use the fact that $\cup \mathcal{T}$ is the torsion subgroup of $\hat{G}$.)

(e) (See [60].) If G is totally disconnected, then $h_a(\Sigma) = h_t(\Sigma) = h(\Sigma^m)$, where m is the Haar measure on G.

(f) Let Σ_0 be the subsystem (G_0, τ). Show that

$$h(\Sigma) = h_a(\Sigma) + h(\Sigma_0).$$

Entropy of Induced Systems

26. Let $\Phi = (X, \mathcal{B}, \mu, \phi)$ be an abstract dynamical system. For each (necessarily countable) partition of X into sets A_n of positive measure, define

$$H(\mathcal{A}) = -\sum_n \mu(A_n) \log \mu(A_n),$$

where $\mathcal{A}$ the smallest σ-algebra containing all the sets A_n. Let $\mathcal{T}$ be the class of all such σ-algebras $\mathcal{A}$ with $H(\mathcal{A}) < \infty$.

(a) Show (e.g., as in Exercise 25) that

$$h(\phi, \mathcal{A}) = \lim_{n\to\infty} \frac{1}{n} H\left(\bigvee_{i=0}^{n-1} \phi^{-i}\mathcal{A}\right)$$

exists and is finite for each $\mathcal{A} \in \mathcal{T}$.

(b) Define $h^*(\Phi) = \sup_{\mathscr{A} \in \mathscr{T}} h(\phi, \mathscr{A})$ and show that $h^*(\Phi) = h(\Phi)$ as previously defined.

(c) If $\mathscr{A}_0 \in \mathscr{T}$ and $\mathscr{T}_0 = \{\mathscr{A} \in \mathscr{T} : \mathscr{A}_0 \subseteq \mathscr{A}\}$, show that

$$h(\Phi) = \sup_{\mathscr{A} \in \mathscr{T}_0} h(\phi, \mathscr{A}).$$

27. (*Kac's theorem*) Let $\Phi = (X, \mathscr{B}, \mu, \phi)$ be an invertible, ergodic system, and let A and n_A be as in Theorem 1.5. Define

$$A_n = \{x \in A : n_A(x) = n\}$$

$$A_{nk} = \phi^k(A_n) \qquad (k = 0, 1, \ldots, n-1; n = 1, 2, \ldots).$$

Show that $\{A_{nk}\}$ is a partition of X, and hence that

$$\int_Y n_A(x)\, \mu(dx) = \mu(X) = 1.$$

Show that 1 is an upper bound on the integral even if Φ is not ergodic. Can the assumption of invertibility also be dropped?

28. Establish the validity of (β) in the proof of Theorem 4.13.

CHAPTER

V

Bernoulli Systems and Ornstein's Theorem

The brief history of ergodic theory has had three high points, each of which evoked a flurry of activity in the field. The first was the proof by G. D. Birkhoff in 1931 of the individual ergodic theorem, which could be said to be the beginning of the mathematical theory. The second was the introduction in 1957 by A. N. Kolmogorov of entropy into ergodic theory. We have already discussed this in Chapter IV. The third was the proof by D. S. Ornstein in 1970 [45, 46] that Bernoulli systems with the same entropy are isomorphic.

A number of other results flow from this basic result and the tools developed to prove it. For example, it is now known that there are Kolmogorov systems (see Exercises I.25 and IV.16) with the same entropy which are not isomorphic to each other or to any Bernoulli systems. Moreover, effective methods have been developed by Ornstein and others to establish whether or not certain classical physical systems are Bernoulli. Many of these results are quite recent and not clearly understood yet.

1. DEFINITIONS

In Chapter I, Example 3, we defined the Bernoulli shift on k points $(1, 2, \ldots, k)$ with distribution $\pi = (p_1, p_2, \ldots, p_k)$. Let us denote this system by $\Phi(\pi)$. It was shown in Chapter III that

$$h(\Phi(\pi)) = -\sum_{j=1}^{k} p_j \log p_j. \tag{1}$$

Thus, clearly, there are distinct distributions π and π' with $h(\Phi(\pi)) = h(\Phi(\pi'))$. The main result of this chapter asserts that such systems are m-isomorphic. Meshalkin [44] had already shown this to be true for certain nonidentical π and π' in 1959. A slightly more general result was given by Blum and Hanson in 1963 [8]. However, the complete solution had to wait for the powerful methods developed by Ornstein in 1970.

Throughout this chapter we shall be concerned with ergodic systems $\Phi = (X, \mathscr{B}, \mu, \phi)$ with positive entropy. Thus it is without loss of generality (Exercise IV.14) that we assume $(X, \mathscr{B}, \mu)$ is nonatomic. We shall also assume throughout that Φ is invertible.

The concepts and definitions to be introduced now are due to Ornstein and were developed specially for the proof of the isomorphism theorem. First, however, we recall and extend the definition of a Bernoulli system.

Let $\pi = (p_1, p_2, p_3, \ldots)$ be a probability distribution on the positive integers Z^+, that is, $p_i \geq 0$ $(i = 1, 2, \ldots)$ and $\sum_i p_i = 1$. In order to include the finite case, we do not insist that all p_i be strictly positive. In the usual way (cf. Example I.3), we define $(X, \mathscr{B}, \mu)$ to be the two-sided direct product of countably many copies of the finite measure space $(Z^+, 2^{Z^+}, \pi)$,

$$(X, \mathscr{B}, \mu) = \mathop{\times}_{n=-\infty}^{\infty} (Z^+, 2^{Z^+}, \pi).$$

The shift σ is defined on X by

$$\sigma(x) = y \qquad \text{with} \quad y_n = x_{n+1},$$

and is an invertible measure-preserving transformation on $(X, \mathscr{B}, \mu)$.

Definition 5.1 The dynamical system $\Phi = (X, \mathscr{B}, \mu, \sigma)$ is called the *Bernoulli system* (or Bernoulli shift) *with distribution* π, and will be denoted $\Phi(\pi)$.

Remark With our understanding that equivalent systems are to be identified, the preceding definition is clearly equivalent to the following one.

Definition 5.1′ $\Phi = (X, \mathscr{B}, \mu, \phi)$ is a *Bernoulli system with distribution* π provided that there exists a countable partition $\mathbb{P} = (P_1, P_2, P_3, \ldots)$ of X such that

(i) each $P_k \in \mathscr{B}$,
(ii) $\mu(P_k) = p_k$ for each k,
(iii) $\bigcup_{n=-\infty}^{\infty} \phi^n(\mathbb{P})$ generates $\mathscr{B}$, and
(iv) $\mu(\bigcap_{j=1}^{l} \phi^{n_j}(P_{k_j})) = \prod_{j=1}^{l} p_{k_j}$

for each choice of l, $k_1, \ldots, k_l$, and $n_1 < n_2 < \cdots < n_l$.

Throughout this chapter a *partition* will mean an *ordered* disjoint sequence of measurable sets with union X. Thus the (finite) partitions $\mathbb{P} = (P, \tilde{P})$ and $\mathbb{P}' = (\tilde{P}, P)$ are to be considered as different partitions. As usual we shall refer to those elements of $\mathbb{P}$ having positive measure as the *atoms* of $\mathbb{P}$. If $\mathbb{P}$ has only a finite number of atoms, it is said to be a *finite partition.*

Associated with each partition $\mathbb{P}$ of X is the σ-algebra $\mathscr{B}(\mathbb{P}) \subseteq \mathscr{B}$ generated by the atoms of $\mathbb{P}$. Note that different partitions may generate the same σ-algebra, and that each *finite* σ-algebra is generated by the partition formed by its atoms.

Definition 5.2 Let $\mathbb{P}^j = (P_1{}^j, P_2{}^j, \ldots, P_{l_j}^j)$ be finite partitions of X for $j = 1, 2, \ldots, n$. Their *join* is the partition $\mathbb{P} = \bigvee_{j=1}^{n} \mathbb{P}^j$ whose elements are the sets

$$P_{k_1 k_2 \cdots k_n} = \bigcap_{j=1}^{n} P_{k_j}^j.$$

The order on $\mathbb{P}$ is lexicographic; that is,

$$\mathbb{P} = (P_{11\cdots 11}, P_{11\cdots 12}, \cdots, P_{11\cdots 1l_n}, P_{11\cdots 21}, \cdots, P_{l_1 l_2 \cdots l_n}).$$

Definition 5.3 If $\mathbb{P}^1$ and $\mathbb{P}^2$ are partitions of X such that each atom of $\mathbb{P}^1$ is a union of atoms of $\mathbb{P}^2$, then we say $\mathbb{P}^2$ *refines* $\mathbb{P}^1$ and write $\mathbb{P}^1 < \mathbb{P}^2$.

Clearly, if $\mathbb{P}^1 < \mathbb{P}^2$, then $\mathscr{B}(\mathbb{P}^1) \subseteq \mathscr{B}(\mathbb{P}^2)$. In any case, $\mathbb{P}^k < \bigvee_{j=1}^{n} \mathbb{P}^j$ for each $k = 1, \ldots, n$.

Definition 5.4 The *distribution* of the partition $\mathbb{P} = (P_1, P_2, \ldots)$ is the sequence

$$d(\mathbb{P}) = (\mu(P_1), \mu(P_2), \ldots).$$

Definition 5.5 Let $\Phi = (X, \mathscr{B}, \mu, \phi)$ be a dynamical system and $\mathbb{P}$ a partition of X. Then $\mathbb{P}$ is a *generator* for Φ if $\mathscr{B}$ is generated by $\bigcup_{n=-\infty}^{\infty} \phi^n(\mathbb{P})$.

Note that this is equivalent to saying that

$$\mathscr{B} = \bigvee_{n=-\infty}^{\infty} \phi^n \mathscr{B}(\mathbb{P}) = \bigvee_{n=1}^{\infty} \bigvee_{k=-n}^{n} \phi^k \mathscr{B}(\mathbb{P}) = \bigvee_{n=1}^{\infty} \bigvee_{k=-\infty}^{n} \phi^k \mathscr{B}(\mathbb{P})$$
$$= \bigvee_{n=1}^{\infty} \phi^n \left(\bigvee_{k=-\infty}^{0} \phi^k \mathscr{B}(\mathbb{P}) \right).$$

A sequence of partitions of $(X, \mathscr{B}, \mu)$ is said to be *independent* if the associated σ-algebras are independent. This is true for the sequence $\phi^n(\mathbb{P})$ iff condition (iv) of Definition 5.1′ is satisfied. In this case, the Borel zero–one law asserts that every set in

$$\bigcap_{n=1}^{\infty} \phi^{-n} \left(\bigvee_{k=-\infty}^{0} \phi^k \mathscr{B}(\mathbb{P}) \right) = \bigcap_{n=1}^{\infty} \bigvee_{k=n}^{\infty} \phi^{-k} \mathscr{B}(\mathbb{P})$$

has μ-measure 0 or 1. Thus the following restatement of Definition 5.1′ implies that every Bernoulli system is equivalent to a Kolmogorov system. (Set $\mathscr{B}_0 = \bigvee_{k=-\infty}^{0} \phi^k \mathscr{B}(\mathbb{P})$ in Exercise I.25.)

Definition 5.1″ $\Phi = (X, \mathscr{B}, \mu, \phi)$ is a *Bernoulli system with distribution* π iff Φ has a generator $\mathbb{P}$ with distribution $d(\mathbb{P}) = \pi$ such that the sequence $\{\phi^n(\mathbb{P})\}$ is independent. We shall say then that $\mathbb{P}$ is an independent generator for Φ.

Remark It is sufficient to show that the $\phi^n(\mathbb{P})$ are independent for $n \geq 1$ (or $n \geq l$), for then independence of the entire bisequence $\phi^n(\mathbb{P})$ $(-\infty < n < +\infty)$ follows from the fact that ϕ is measure preserving.

Definition 5.6 By a *stack* of height n we shall mean a disjoint sequence $F_0, \ldots, F_{n-1}$ of measurable subsets of the measure space $(X, \mathscr{B}, \mu)$ and a sequence of m-isomorphisms ϕ_j of F_j onto F_{j+1} $(j = 0, 1, \ldots, n-2)$.

If we define ϕ to be ϕ_j on F_j, then ϕ is a measure-preserving transformation of $\bigcup_{j=0}^{n-2} F_j$ onto $\bigcup_{j=1}^{n-1} F_j$, and $F_j = \phi^j F_0$. We can think of the sets F_j lying above one another in a "stack" and ϕ mapping each point in F_j into the one directly above it in F_{j+1} (Fig. 5.1).

Now suppose that $\mathbb{P} = (P_1, P_2, \ldots, P_l)$ is a partition of X. Then $\mathbb{P}$ induces a partition on $\bigcup_{j=0}^{n-1} F_j$ (Fig. 5.2). Moreover, ϕ^{-j} maps F_j onto

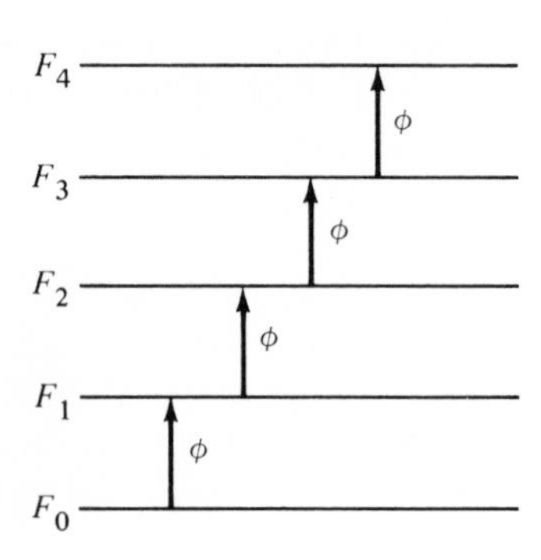

Figure 5.1. Stack: $n = 5$.

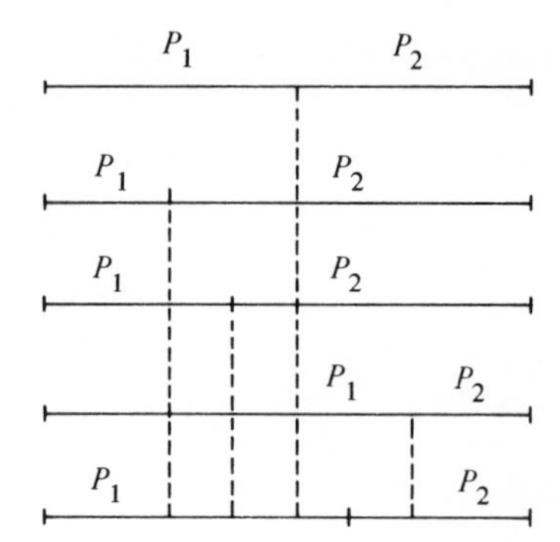

Figure 5.2. Stack and partition: $n = 5$, $l = 2$.

F_0 and provides another partition $\mathbb{P}_j$ of F_0 $(j = 0, 1, \ldots, n-1)$. The join $\mathbb{P}' = \mathbb{P}_0 \vee \mathbb{P}_1 \vee \cdots \vee \mathbb{P}_{n-1}$ is a partition of F_0 into points with identical "futures" with regard to the transformation ϕ and the partition $\mathbb{P}$.

Definition 5.7 The $\mathbb{P}$-*n*-*name* of a point $x \in F_0$ is the sequence $i_0, i_1, \ldots, i_{n-1}$ such that $\phi^j(x) \in P_{i_j}$ $(j = 0, 1, \ldots, n-1)$.

Definition 5.8 If $\{\phi^j F_0 : j = 0, \ldots, n-1\}$ is a stack and $\mathbb{P}$ a partition, a *column* associated with $\mathbb{P}$ is a substack $\{\phi^j A : j = 0, \ldots, n-1\}$, where A is an atom of the partition $\mathbb{P}'$ of F_0 into points with identical $\mathbb{P}$-n-names. Each of the sets $\phi^j A$ is called a *column level* (Fig. 5.3).

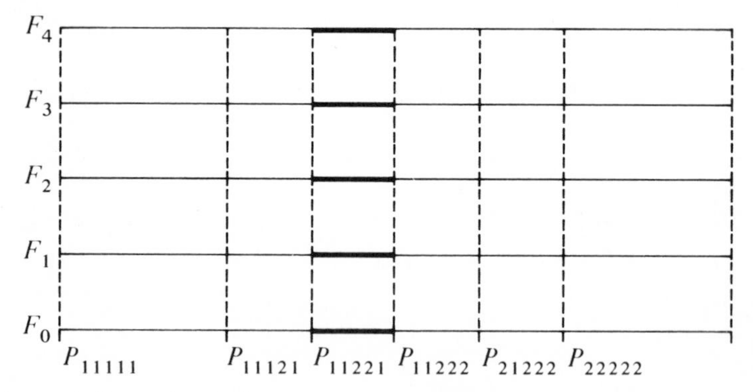

Figure 5.3. Heavy lines indicate the column of points with $\mathbb{P}$-n-name (1, 1, 2, 2, 1).

Clearly, the collection of column levels associated with $\mathbb{P}$ provides a partition $\mathbb{P}''$ of $\bigcup_{j=0}^{n-1} \phi^j F_0$ which is finer than the one induced by $\mathbb{P}$. It is the coarsest partition of the stack which is finer than $\mathbb{P}$ and is "consistent" with ϕ in the sense that ϕ maps atoms of $\mathbb{P}''$ onto atoms of $\mathbb{P}''$.

Definition 5.9 A *gadget* is a quadruple $(F, \phi, n, \mathbb{P})$ such that $F, \phi F, \ldots, \phi^{n-1}F$ forms a stack and $\mathbb{P}$ is a partition of $\bigcup_{j=0}^{n-1} \phi^j F$. The columns associated with $\mathbb{P}$ are called *columns of the gadget*. The *$\mathbb{P}$-n-name* of the column $A, \phi A, \ldots, \phi^{n-1}A$ is the $\mathbb{P}$-n-name of the points $x \in A$.

Remark An equivalent definition of a gadget could be given as a union of disjoint stacks of height n (the columns of the gadget) with distinct assignments of sequences $(j_0, j_1, \ldots, j_{n-1})$ (the $\mathbb{P}$-n-name) from the set $\{1, 2, \ldots, l\}$ to the columns. For then the partition $\mathbb{P}$ may be recovered as follows. The atom P_i of $\mathbb{P}$ is the union of the columns levels $\phi^k A$ such that $j_k = i$ in the sequence assigned to that column.

Clearly, distinct gadgets may give rise to similar appearing column structures. For example, the partitions $\mathbb{P}_1$ and $\mathbb{P}_2$ may contain a different number of sets, or may have the same sets arranged in a different order, and lead to column-level partitions $\mathbb{P}_1''$ and $\mathbb{P}_2''$ which are indistinguishable except for labeling (Fig. 5.4). We shall want to preserve these distinctions, for example, through the recitation of column names. However, we do want to identify gadgets which differ only by a scaling factor. We are thus led to the following definition of gadget isomorphism.

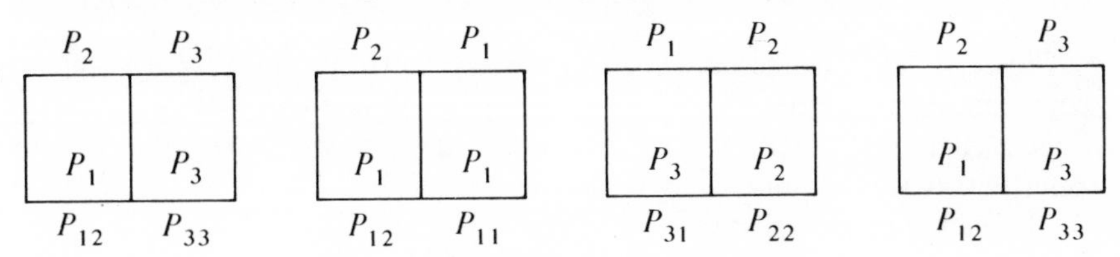

Figure 5.4. Pairwise nonisomorphic gadgets.

Definition 5.10 Two gadgets $\Gamma_1 = (F_1, \phi_1, n, \mathbb{P}_1)$ and $\Gamma_2 = (F_2, \phi_2, n, \mathbb{P}_2)$ are *isomorphic* if they have the same collection of column names and the distributions of the partitions $\mathbb{P}_1'$ and $\mathbb{P}_2'$ of F_1 and F_2, respectively, into column levels agree up to a constant multiple.

Remark To say that Γ_1 and Γ_2 are isomorphic is to say (assuming that F_1 and F_2 are measure-theoretically isomorphic) that there exists an invertible measurable mapping of F_1 onto F_2 which preserves relative "lengths" and the structure of the gadget built on F_1. We shall call such a mapping a *gadget isomorphism*.

In the approximation arguments of subsequent sections we shall want to speak of closeness of partitions or of gadgets and of approximate independence. We are thus led to the following definitions.

Definition 5.11 Let $\mathbb{P} = (P_1, P_2, \ldots)$ and $\mathbb{Q} = (Q_1, Q_2, \ldots)$ be partitions of the same or different measure spaces $(X_1, \mathscr{B}_1, \mu_1)$ and $(X_2, \mathscr{B}_2, \mu_2)$. The *distribution distance* of $\mathbb{P}$ and $\mathbb{Q}$ is

$$\delta(\mathbb{P}, \mathbb{Q}) = \sum_i |\mu_1(P_i) - \mu_2(Q_i)|.$$

Definition 5.12 Let $\mathbb{P} = (P_1, P_2, \ldots)$ and $\mathbb{Q} = (Q_1, Q_2, \ldots)$ be partitions of the same measure space $(X, \mathscr{B}, \mu)$. The *partition distance* of $\mathbb{P}$ and $\mathbb{Q}$ is

$$\rho(P, Q) = \sum_i \mu(P_i \,\Delta\, Q_i).$$

Definition 5.13 Let $\Gamma_1 = (F_1, \phi_1, n, \mathbb{P}_1)$ and $\Gamma_2 = (F_2, \phi_2, n, \mathbb{P}_2)$ be gadgets. Let Ψ be the collection of invertible mappings ψ of F_2 onto F_1 which are measure preserving up to a constant factor. The *gadget distance* of Γ_1 and Γ_2 is

$$\gamma(\Gamma_1, \Gamma_2) = \inf_{\psi \in \Psi} \frac{1}{n} \sum_{i=0}^{n-1} \rho(\phi_1^{-i}\mathbb{P}_1 \cap F_1, \psi(\phi_2^{-i}\mathbb{P}_2 \cap F_2)),$$

where ρ is the partition distance on F_1. If $\Psi = \varnothing$, set $\gamma(\Gamma_1, \Gamma_2) = +\infty$.

Closely related to the gadget distance is the process distance associated with pairs $(\Phi, \mathbb{P})$ where $\Phi = (X, \mathscr{B}, \mu, \phi)$ is a dynamical system and $\mathbb{P}$ a finite measurable partition of X. The name arises from the fact that $(\Phi, \mathbb{P})$ determines a (finite-valued) stochastic process $\{f_n\}$ by $f_n(x) = i$ if $\phi^n x \in P_i \in \mathbb{P}$. To define the process distance, we consider measure isomorphisms ψ of X_2 onto X_1, as we did isomorphisms of F_2 onto F_1 for the gadget distance. Let $\Psi(X_2, X_1)$ denote the class of such isomorphisms.

Definition 5.14 Let $\Phi_i = (X_i, \mathscr{B}_i, \mu_i, \phi_i)$ be dynamical systems and $\mathbb{P}_i$ finite, measurable partitions of X_i with the same number of atoms $(i = 1, 2)$. The *process distance* is defined by

$$\pi((\Phi_1, \mathbb{P}_1), (\Phi_2, \mathbb{P}_2)) = \sup_n \inf_{\psi \in \Psi(X_2, X_1)} \frac{1}{n} \sum_{i=0}^{n-1} \rho(\phi_1^{-i}\mathbb{P}_1, \psi\phi_2^{-i}\mathbb{P}_2).$$

Our final definition of this section gives a quantitative interpretation to the statement that two partitions $\mathbb{P}$ and $\mathbb{Q}$ are almost independent. It says that $\mathbb{P}$ partitions almost all of the sets in $\mathbb{Q}$ in almost the same proportions as it partitions X.

If Q is a measurable subset of X with $0 < \mu(Q) < \infty$, we denote by $\mathbb{P} \cap Q$ the partition $\{P \cap Q : P \in \mathbb{P}\}$ of the measure space $(Q, Q \cap \mathscr{B}, (1/\mu(Q))\mu)$.

Definition 5.15 Let $\mathbb{P}$ and $\mathbb{Q}$ be partitions of $(X, \mathscr{B}, \mu)$, and let $\varepsilon > 0$. Then $\mathbb{P}$ is *ε-independent* of $\mathbb{Q}$ if $\mathbb{Q}$ is the disjoint union of collections $\mathbb{Q}_1$ and $\mathbb{Q}_2$ with

(i) $\mu(\cup \mathbb{Q}_2) < \varepsilon$, and
(ii) $\delta(\mathbb{P}, \mathbb{P} \cap Q) < \varepsilon$ for all $Q \in \mathbb{Q}_1$.

Remarks *1* The function δ is a pseudometric on partitions of finite measure spaces. In fact, $\delta(\mathbb{P}, \mathbb{Q}) = 0$ iff $\mathbb{P}$ and $\mathbb{Q}$ have the same distribution, and δ is the l_1-metric on the sequences $d(\mathbb{P})$. In particular, the collection of partitions of a given finite measure space is *complete* in the δ-topology.

2 The function ρ is a metric on the space of partitions of the finite measure space $(X, \mathscr{B}, \mu)$. Moreover,

$$\delta(\mathbb{P}, \mathbb{Q}) \leq \rho(\mathbb{P}, \mathbb{Q}).$$

3 The function γ is a pseudometric. Assuming that F_1 and F_2 are measure-theoretically isomorphic modulo a constant multiplicative factor, $\gamma(\Gamma_1, \Gamma_2)$ measures how well an isomorphic copy of the gadget Γ_2 on the space of the gadget Γ_1 can be made to fit Γ_1 at every level of Γ_1. In particular, $\gamma(\Gamma_1, \Gamma_2) = 0$ iff there is a gadget isomorphism of Γ_1 and Γ_2.

4 Each partition $\mathbb{P}$ of X determines a factor $\Phi_{\mathscr{B}(\mathbb{P})}$ as explained in the previous chapter. Thus the pseudometric π may be thought of as a measure of closeness of $\Phi_{\mathscr{B}(\mathbb{P}_1)}$ and $\Phi_{\mathscr{B}(\mathbb{P}_2)}$. In fact, it is not hard to see that $\pi((\Phi_1, \mathbb{P}_1), (\Phi_2, \mathbb{P}_2)) = 0$ iff $\Phi_{\mathscr{B}(\mathbb{P}_1)} \cong \Phi_{\mathscr{B}(\mathbb{P}_2)}$. In case $\mathbb{P}_1$ and $\mathbb{P}_2$ are generators for Φ_1 and Φ_2, respectively, π gives a useful measure of "almost-isomorphism" of Φ_1 and Φ_2.

5 The usual analytic trickery of comparing the nrth term to the nth term of the sequence shows that, in fact, $\sup_n$ can be replaced by $\lim_{n\to\infty}$ in the definition of π.

6 It is reasonable, and not hard to show for ergodic systems, that closeness in the process metric is equivalent to closeness in the gadget metric for arbitrarily long gadgets. This requires the use of Lemma 5.3 below.

7 Partitions $\mathbb{P}$ and $\mathbb{Q}$ of X are independent iff $\mathbb{P}$ is ε-independent of $\mathbb{Q}$ for each $\varepsilon > 0$.

8 If $\mathbb{P}$ is $\varepsilon/2$-independent of $\mathbb{Q}$, then

(a) $\sum_i \sum_j |\mu(P_i \cap Q_j) - \mu(P_i)\, \mu(Q_j)| < \varepsilon$.

Conversely, if (a) holds, then $\mathbb{P}$ is $\sqrt{\varepsilon}$-independent of $\mathbb{Q}$. Since (a) is symmetric in $\mathbb{P}$ and $\mathbb{Q}$, it follows that $\mathbb{P}$ ε-independent of $\mathbb{Q}$ implies that $\mathbb{Q}$ is $\sqrt{2\varepsilon}$-independent of $\mathbb{P}$.

9 If $\mathbb{P}$ is ε-independent of $\mathbb{Q}$, then there exists a measurable set A, with $\mu(X \sim A) < 2\varepsilon$, such that $\mathbb{P} \cap A$ and $\mathbb{Q} \cap A$ are independent.

2. APPROXIMATION LEMMAS

The general outline of the proof of the ismorphism theorem as well as the proofs of several of the lemmas follow Shields [57].

We shall assume throughout this and the following sections that all measure spaces are normalized, that is, have total measure one, and all partitions are finite.

For ease of later reference, we shall begin by stating all of the lemmas to be proved in this section. The first six lemmas involve only the concepts developed in the previous section.

Lemma 5.1 (*Law of large numbers*) *Let $\Phi = (X, \mathscr{B}, \mu, \phi)$ be an ergodic system, let $\mathbb{P}$ be a partition of X, and let $\varepsilon > 0$. For all sufficiently large n there is a set $C_n \in \bigvee_{i=0}^{n-1} \mathscr{B}(\phi^{-i}\mathbb{P})$ such that*

(i) $\mu(X \sim C_n) < \varepsilon$, *and*
(ii) $|(1/n) \sum_{i=0}^{n-1} \chi_P(\phi^i x) - \mu(P)| < \varepsilon$ *for all* $P \in \mathbb{P}$, $x \in C_n$.

Lemma 5.2 (*Rohlin*) *Let $\Phi = (X, \mathscr{B}, \mu, \phi)$ be an ergodic dynamical system, and let $\varepsilon > 0$. Then for each n there exists a set $F \in \mathscr{B}$ such that*

(i) $\phi^i F \cap \phi^j F = \varnothing$ $(i \neq j;\ i, j = 0, 1, \ldots, n-1)$, *and*
(ii) $\mu(\bigcup_{i=0}^{n-1} \phi^i F) > 1 - \varepsilon$.

Lemma 5.3 *Let Φ and ε be as in Lemma 5.2, and let $\mathbb{P}$ be a finite partition of X. Then for each n there exists a set $F \in \mathscr{B}$ satisfying* (i) *and* (ii) *of that lemma and*

(iii) $\delta(\mathbb{P}, \mathbb{P} \cap F) < 2\varepsilon$.

Lemma 5.4 *Let $\Gamma = (F, \phi, n, \mathbb{P})$ be a gadget in $(X, \mathscr{B}, \mu)$, let $\Phi' = (X', \mathscr{B}', \mu', \phi')$ be an ergodic dynamical system, and let $\varepsilon > 0$. Then there exists a set $F' \in \mathscr{B}'$ and a finite partition $\mathbb{P}'$ of X' such that Γ and $\Gamma' = (F', \phi', n, \mathbb{P}')$ are isomorphic gadgets, and*

$$\mu\left(\bigcup_{i=0}^{n-1} \phi'^i F'\right) > 1 - \varepsilon.$$

Lemma 5.5 *Suppose $(F, \phi, n, \mathbb{P})$ and $(F', \phi', n, \mathbb{P}')$ are isomorphic gadgets. Let $\mathbb{Q}$ be a partition of $\bigcup_{i=0}^{n-1} \phi^i F$. Then there is a partition $\mathbb{Q}'$ of $\bigcup_{i=0}^{n-1} \phi'^i F'$ such that $(F, \phi, n, \mathbb{P} \vee \mathbb{Q})$ is isomorphic to $(F', \phi', n, \mathbb{P}' \vee \mathbb{Q}')$.*

Lemma 5.6 *Let $\Gamma = (F, \phi, n, \mathbb{P})$ and $\Gamma' = (F', \phi', n, \mathbb{P}')$ be gadgets, and let $\varepsilon > 0$. Suppose that*

(a) $d(\bigvee_{i=0}^{n-1} \phi^{-i}\mathbb{P} \cap F) = d(\bigvee_{i=0}^{n-1} \phi^{-i}\mathbb{P})$,
(b) $d(\bigvee_{i=0}^{n-1} \phi'^{-1}\mathbb{P}' \cap F') = d(\bigvee_{i=0}^{n-1} \phi'^{-i}\mathbb{P}')$,
(c) $\{\phi^{-i}\mathbb{P} \cap F\}$ *is an* ε*-independent sequence,*
(d) $\{\phi'^{-i}\mathbb{P}' \cap F'\}$ *is an independent sequence, and*
(e) $\delta(\mathbb{P}, \mathbb{P}') < \varepsilon$.

Then

$$\gamma(\Gamma, \Gamma') < 3\varepsilon.$$

The remaining lemmas involve also the notion of entropy (Chapter IV). For each dynamical system $\Phi = (X, \mathscr{B}, \mu, \phi)$ and each (finite) partition $\mathbb{P}$ of X, we shall denote by $h(\phi, \mathbb{P})$ the entropy $h(\phi, \mathscr{B}(\mathbb{P}))$ of ϕ on the σ-algebra $\mathscr{B}(\mathbb{P})$ generated by $\mathbb{P}$ (Definition 4.2). Likewise, we write $H(\mathbb{P}) = H(\mathscr{B}(\mathbb{P}))$.

Lemma 5.7 (*McMillan*) *Let $\Phi = (X, \mathscr{B}, \mu, \phi)$ be an ergodic system, let $\mathbb{P}$ be a partition of X, and let $\varepsilon > 0$. For all sufficiently large n there is a collection $\mathscr{C}_n$ of atoms of $\bigvee_{i=0}^{n-1} \phi^{-i}\mathbb{P}$ such that*

(i) $\mu(\cup \mathscr{C}_n) \geq 1 - \varepsilon$, *and*
(ii) $|h(\phi, \mathbb{P}) + (1/n) \log \mu(A)| < \varepsilon$ *for* $A \in \mathscr{C}_n$.

Consequently, the number $\nu(\mathscr{C}_n)$ of atoms in $\mathscr{C}_n$ satisfies

(iii) $(1 - \varepsilon)e^{n(h(\phi, \mathbb{P}) - \varepsilon)} \leq \nu(\mathscr{C}_n) \leq e^{n(h(\phi, \mathbb{P}) + \varepsilon)}$.

Lemma 5.8 *Let Φ be a dynamical system, let k be a positive integer greater than 2, and let $\varepsilon > 0$. Then there exists a $\delta = \delta(\varepsilon, k) > 0$ such that, whenever $\mathbb{P}$ is a partition containing k atoms and satisfying*

$$h(\phi, \mathbb{P}) \geq H(\mathbb{P}) - \delta,$$

it follows that $\{\phi^n\mathbb{P}\}$ is an ε-independent sequence. In particular,

$$h(\phi, \mathbb{P}) = H(\mathbb{P})$$

iff $\{\phi^n\mathbb{P}\}$ is independent.

Lemma 5.9 *Let Φ be a Bernoulli system with independent generator $\mathbb{P}$, and let $\varepsilon > 0$. Then there exists a $\delta > 0$ such that, whenever Φ' is ergodic and $\mathbb{P}'$ is a partition of X' with the same number of atoms as $\mathbb{P}$ such that*

(i) $\delta(\mathbb{P}, \mathbb{P}') < \delta$, *and*
(ii) $|H(\mathbb{P}) - h(\phi', \mathbb{P}')| < \delta$,

we have

(iii) $\pi((\Phi, \mathbb{P}), (\Phi', \mathbb{P}')) < \varepsilon$.

Lemma 5.9 turns out to be a "good" theorem and like all good theorems has become a definition.

Definition 5.16 Let Φ be an ergodic system, and let $\mathbb{P}$ be a finite partition of X. Then $\mathbb{P}$ is said to be *finitely determined* (relative to Φ) if given $\varepsilon > 0$ there is a positive integer n and a $\delta > 0$ such that, whenever Φ' is ergodic and $\mathbb{P}'$ is a partition of X' such that

(i) $h(\Phi') \geq h(\Phi, \mathbb{P})$,
(ii) $\nu(\mathbb{P}') = \nu(\mathbb{P})$,
(iii) $\delta(\bigvee_{i=0}^{n-1} \phi'^{-i}\mathbb{P}', \bigvee_{i=0}^{n-1} \phi^{-i}\mathbb{P}) < \delta$, *and*
(iv) $0 \leq h(\phi, \mathbb{P}) - h(\phi', \mathbb{P}') < \delta$,

then

(v) $\pi((\Phi, \mathbb{P}), (\Phi', \mathbb{P}')) < \varepsilon$.

Remark It is easily seen from Lemma 5.9 that independent generators are finitely determined. The importance of finitely determined partitions lies in the (not obvious) facts that they are precisely the ones that generate Bernoulli systems and are often easier to find than independent generators.

Lemma 5.10 *Let Φ be a Bernoulli system with independent generator $\mathbb{P}$. Let Φ' be any ergodic dynamical system with $h(\Phi') \geq h(\Phi)$, and let $\varepsilon > 0$. Then there exists a partition $\mathbb{Q}$ such that*

(i) $\nu(\mathbb{Q}) = \nu(\mathbb{P})$,
(ii) $\delta(\mathbb{P}, \mathbb{Q}) < \varepsilon$, *and*
(iii) $0 \leq h(\phi, \mathbb{P}) - h(\phi', \mathbb{Q}) < \varepsilon$.

Lemma 5.11 *Let Φ be an ergodic system, let $\mathbb{P}$ be a finitely determined partition of X, and let $\varepsilon > 0$. Then there exists a positive integer n_1 and a $\delta_1 > 0$ such that, whenever Φ' is an ergodic system and $\mathbb{P}'$ is a partition of X' such that*

(i) $h(\Phi') \geq h(\Phi, \mathbb{P})$,
(ii) $\nu(\mathbb{P}') = \nu(\mathbb{P})$,
(iii) $\delta(\bigvee_{i=0}^{n_1-1} \phi'^{-i}\mathbb{P}', \bigvee_{i=0}^{n_1-1} \phi^{-i}\mathbb{P}) < \delta_1$, *and*
(iv) $0 \leq h(\phi, \mathbb{P}) - h(\phi', \mathbb{P}') < \delta_1$,

then for any $\delta_2 > 0$ and n_2 there is a partition $\mathbb{Q}$ such that

(v) $\delta(\bigvee_{i=0}^{n_2-1} \phi'^{-1}\mathbb{Q}, \bigvee_{i=0}^{n_2-1} \phi^{-i}\mathbb{P}) < \delta_2$,
(vi) $0 \leq h(\phi', \mathbb{Q}) - h(\phi, \mathbb{P}) < \delta_2$, *and*
(vii) $\rho(\mathbb{P}, \mathbb{Q}) < \varepsilon$.

Lemma 5.12 (*Ornstein's principal lemma*) *Let Φ be a Bernoulli system with independent generator $\mathbb{P}$, and let $\varepsilon > 0$. Then there exists a $\delta > 0$ such that, whenever Φ' is ergodic and $\mathbb{P}'$ is a partition of X' satisfying*

(i) $h(\Phi') \geq h(\Phi)$,
(ii) $v(\mathbb{P}') = v(\mathbb{P})$,
(iii) $\delta(\mathbb{P}, \mathbb{P}') < \delta$, *and*
(iv) $0 \leq h(\phi, \mathbb{P}) - h(\phi', \mathbb{P}') < \delta$,

there is a partition $\mathbb{Q}$ such that

(v) $\{\phi^n\mathbb{Q}\}$ *is an independent sequence,*
(vi) $d(\mathbb{Q}) = d(\mathbb{P})$, *and*
(vii) $\rho(\mathbb{P}, \mathbb{Q}) < \varepsilon$.

Proof (Lemma 5.1) According to Proposition 1.4, the quantity on the left in (ii) converges to zero as $n \to \infty$ for almost all $x \in X$ and each $P \in \mathbb{P}$. It follows that

$$\lim_{n\to\infty} \mu\left\{x \in X: \left|\frac{1}{n}\sum_{i=0}^{n-1} \chi_P(\phi^i x) - \mu(P)\right| \geq \varepsilon \text{ for some } P \in \mathbb{P}\right\} = 0.$$

Let D_n be the indicated set, so that $\mu(D_n) \to 0$. Clearly, $D_n \in \bigvee_{i=0}^{n-1} \mathscr{B}(\phi^{-i}\mathbb{P})$. Choose N so that $n \geq N \Rightarrow \mu(D_n) < \varepsilon$, and set $C_n = X \sim D_n$. ∎

Remark If $\mathbb{P} = \{P_1, \ldots, P_l\}$, then the quantity

$$f(x, j, n) = \frac{1}{n}\sum_{i=0}^{n-1} \chi_{P_j}(\phi^i x)$$

is the relative frequency of occurrence of j in the $\mathbb{P}$-n-name of x. This is constant on the atoms of $\bigvee_{i=0}^{n-1} \phi^{-i}\mathbb{P} = \mathbb{P}_n$. Thus Lemma 5.1 says that for large n the relative frequency of occurrence of j in the $\mathbb{P}$-n-name of the atom $A \in \mathbb{P}_n$ is near $\mu(P_j)$ for most of the atoms of $\mathbb{P}_n$.

Proof (Lemma 5.2) Our proof, the idea of which is due to Ornstein, follows [32, pp. 70–72], where Lemma 5.2 is proved for antiperiodic Φ, that is, under the assumption that $\phi^n x \neq x$ for all $n \geq 1$ and almost all $x \in X$. It is easily seen (Exercise 1) that Φ ergodic with positive entropy implies Φ is antiperiodic.

Let p be a positive integer such that $p^{-1} < \varepsilon$. We begin by constructing a measurable set F with $\mu(F) > 0$ such that

(i) $F, \phi F, \phi^2 F, \ldots, \phi^{pn-1}F$ are pairwise disjoint, and
(ii) $G, \phi G, \phi^2 G, \ldots, \phi^{pn-1}G$ pairwise disjoint with

$$F \subseteq G \Rightarrow \mu(G \sim F) = 0.$$

First let $E_1 \in \mathscr{B}$ satisfy $\mu(E_1) > 0$ and $\mu(E_1 \,\Delta\, \phi E_1) > 0$ (by ergodicity). Let $F_1 = E_1 \sim \phi E_1$. Since $\mu(E_1 \sim \phi E_1) = \mu(\phi E_1 \sim E_1) = \frac{1}{2}\mu(E_1 \,\Delta\, \phi E_1)$, it follows that $\mu(F_1) > 0$. Moreover, $F_1 \cap \phi F_1 = \varnothing$. Having chosen $F_1 \supseteq F_2 \supseteq \cdots \supseteq F_k$ with $\mu(F_k) > 0$ and $F_k, \phi F_k, \ldots, \phi^k F_k$ pairwise disjoint, there must exist $E_{k+1} \subseteq F_k$ with $\mu(E_{k+1}) > 0$ and $\mu(E_{k+1} \,\Delta\, \phi^{k+1} E_{k+1}) > 0$. In fact, any subset E of F_k with $\mu(E) > 0$ and $\mu(F_k \sim E) > 0$ will do. For otherwise, $A = E \cup \phi E \cup \cdots \cup \phi^k E$ would be an invariant set with $\mu(A) > 0$ and $\mu(X \sim A) > 0$. As before, let $F_{k+1} = E_{k+1} \sim \phi^{k+1} E_{k+1}$, so that $F_{k+1} \cap \phi^{k+1} F_{k+1} = \varnothing$. Since clearly $\phi^i F_{k+1} \cap \phi^j F_{k+1} = \varnothing$ for $0 \le i < j \le k+1$, we have by induction defined for each k a set F_k with $\mu(F_k) > 0$ and $F_k, \phi F_k, \ldots, \phi^k F_k$ pairwise disjoint.

Setting $k = pn - 1$, it follows that the class $\mathscr{F}$ of sets $F \in \mathscr{B}$ with $\mu(F) > 0$ and satisfying (i) above is nonempty. Identifying sets in $\mathscr{F}$ which differ by a set of measure zero and Zornifying, we obtain (ii).

Now if F_0 is a measurable subset of $\phi^{pn-1}F$ of positive measure, then $\mu(\phi^j F_0 \cap F) > 0$ for some $j = 1, \ldots, pn$. For otherwise, $F \cup \phi F_0 \in \mathscr{F}$ contradicting (ii). Set

$$A_j = \phi^{pn-1}F \cap \left[\phi^{-j}F \sim \sum_{i=1}^{j-1} \phi^{-i}F\right] \qquad (1 \le j \le pn).$$

Then the A_j ($j = 1, \ldots, pn$) are pairwise disjoint and

$$\mu\left(\phi^{pn-1}F \sim \bigcup_{j=1}^{pn} A_j\right) = 0.$$

Consider the sets $\phi^i A_j$ for $1 \le i < j \le pn$. For fixed i they are transforms by ϕ^i of pairwise disjoint sets, hence pairwise disjoint. For $i_1 < i_2$, $\phi^{-i_2}(\phi^{i_2} A_{j_2}) = A_{j_2} \subseteq \phi^{pn-1}F$ and $\phi^{-i_2}(\phi^{i_1} A_{j_1}) \subseteq \phi^{pn-1+i_1-i_2}F$ are disjoint since $i_2 - i_1 < pn - 1$. Thus the total collection of $\phi^i A_j$ is pairwise disjoint. Moreover, each of them is disjoint from each $\phi^k F$ $(0 \le k \le pn-1)$, as is easily seen by considering separately the cases $k > i$ (since $\phi^{pn-1}F \cap \phi^{k-i}F = \varnothing$) and $k \le i$ (since $i - k < j$ and so $\phi^{i-k}A_j \cap F = \varnothing$).

In particular, $\phi A_1, \phi^2 A_2, \ldots, \phi^{pn} A_{pn}$ are pairwise disjoint subsets of F.

For if $x = \phi^i y = \phi^{i+j} z$ with $y \in A_i$ and $z \in A_{i+j}$, then $y = \phi^j z \in A_i \cap \phi^j A_{i+j} \subseteq F \cap \phi^j A_{i+j}$. Moreover,

$$\sum_{j=1}^{pn} \mu(\phi^j A_j) = \sum_{j=1}^{pn} \mu(A_j) = \mu(\phi^{pn-1} F) = \mu(F).$$

Setting

$$F^* = \left(\bigcup_{k=0}^{pn-1} \phi^k F\right) \cup \left(\bigcup_{1 \le i < j \le pn} \phi^i A_j\right),$$

we see that $\mu(F^* \, \Delta \, \phi F^*) = 0$. By ergodicity, it follows that $\mu(F^*) = 1$.

Finally, setting

$$E = \left(\bigcup_{k=0}^{p-1} \phi^{kn} F\right) \cup \left(\bigcup_{i+n-1<j} \phi^i A_j\right),$$

we see that E, ϕE, $\ldots$, $\phi^{n-1} E$ are pairwise disjoint. Their union differs from F^* by the union of a collection of sets $\phi^i A_j$, consisting of less than n of these sets for each j. Thus

$$\mu\left(F^* \sim \bigcup_{i=0}^{n-1} \phi^i E\right) \le \sum_{j=1}^{pn} n\mu(A_j) = n\mu(F).$$

On the other hand, since F, ϕF, $\ldots$, $\phi^{pn-1} F$ are pairwise disjoint, $n\mu(F) \le n \cdot 1/pn = 1/p < \varepsilon$, and the proof is complete. ▮

Proof (Lemma 5.3) Recall that $\mathbb{P} \cap F$ is the partition of F given by $\mathbb{P} \cap F = \{A \cap F : A \in \mathbb{P}\}$ and its distribution is given by the *normalized* measure induced on F by μ. Thus condition (iii) says that

$$\sum_{A \in \mathbb{P}} \left| \mu(A) - \frac{\mu(A \cap F)}{\mu(F)} \right| < 2\varepsilon.$$

Now use Lemma 5.2 to obtain a stack F_0, ϕF_0, $\ldots$, $\phi^{m-1} F_0$ such that (a) $m = np - 1$, (b) $\mu(F_0) < \varepsilon/2n$, and (c) $\mu(\bigcup_{i=0}^{m-1} \phi^i F_0) > 1 - \varepsilon/2$. The partition $\mathbb{P}$ induces a partition $\mathbb{P}' = \bigvee_{i=0}^{m-1} (\phi^{-i}\mathbb{P}) \cap F_0$ on F_0 corresponding to the columns of the gadget $(F_0, \phi, m, \mathbb{P})$. Since X is nonatomic, we can partition each atom $A \in \mathbb{P}'$ into n pieces of equal measure, $A = \bigcup_{j=0}^{n-1} A_j$. Let

$$F = \bigcup_{A \in \mathbb{P}'} \bigcup_{j=0}^{n-1} \bigcup_{i=0}^{p-2} \phi^{ni+j} A_j.$$

(See Fig. 5.5.)

Clearly, F, ϕF, $\ldots$, $\phi^{n-1} F$ are pairwise disjoint, and their union includes all of $\bigcup_{i=0}^{m-1} \phi^i F_0$ except a set of measure $(n-1)\mu(F_0) < \varepsilon/2$, which proves (ii).

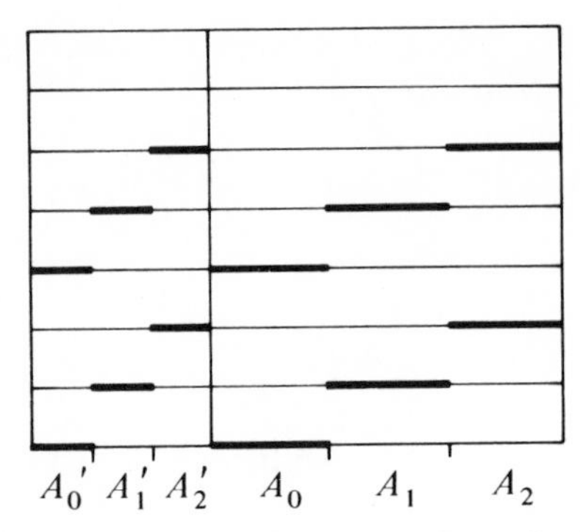

***Figure* 5.5.** $n = 3$, $p = 3$, $m = 8$. Two atoms. Heavy line indicates F.

In addition, we have $d(\mathbb{P} \cap F) = d(\mathbb{P} \cap (\bigcup_{i=0}^{m-n} \phi^i F_0))$, since the atoms of the latter partition consist of unions of column levels of the gadget $(F_0, \phi, m, \mathbb{P})$ and we have assigned exactly $1/n$ of the measure of each column level to F.

On the other hand, $\bigcup_{i=0}^{m-n} \phi^i F_0 = H$ has measure no less than $1 - \varepsilon/2 - n\mu(F_0) > 1 - \varepsilon$. Thus

$$\begin{aligned}\delta(\mathbb{P}, \mathbb{P} \cap H) &= \sum_{A \in \mathbb{P}} \left| \mu(A) - \frac{\mu(A \cap H)}{\mu(H)} \right| \\ &\leq \sum_{A \in \mathbb{P}} [\mu(A) - \mu(A \cap H] + \sum_{A \in \mathbb{P}} \left[\frac{\mu(A \cap H)}{\mu(H)} - \mu(A \cap H) \right] \\ &= 2(1 - \mu(H)) < 2\varepsilon. \ \blacksquare\end{aligned}$$

Remark The last part of the proof shows that whenever H is a measurable set with $\mu(H) > 1 - \varepsilon$ and $\mathbb{P}$ is a partition we have $\delta(\mathbb{P}, \mathbb{P} \cap H) < 2\varepsilon$.

Proof (Lemma 5.4) From Lemma 5.2, let $F' \in \mathscr{B}'$ be chosen so that $F', \phi' F', \ldots, \phi'^{n-1} F'$ is a disjoint sequence, and $\mu'(\bigcup_{i=0}^{n-1} \phi'^i F') > 1 - \varepsilon$. Let $\mathbb{Q}'$ be any partition of F' such that

$$d(\mathbb{Q}') = d\left(\left(\bigvee_{i=0}^{n-1} \phi^{-i}\mathbb{P}\right) \cap F\right). \tag{2}$$

Then $\mathbb{Q}'$ determines a partition of $\bigcup_{i=0}^{n-1} \phi'^i F'$ into column levels $\phi'^i A$, $A \in \mathbb{Q}'$. Clearly, (2) determines an assignment of column names to the stack over F_0, and thus defines a partition $\mathbb{P}'$ of that stack (see the Remark following Definition 5.9) such that the gadgets Γ and $\Gamma' = (F', \phi', n, \mathbb{P}')$ are isomorphic. $\blacksquare$

Proof (Lemma 5.5) Again we can choose Q_0' to be any partition of F' satisfying

$$d(\mathbb{Q}_0') = d\left(\left(\bigvee_{i=0}^{n-1} \phi^{-i}(\mathbb{P} \vee \mathbb{Q})\right) \cap F\right) \tag{3}$$

and

$$\mathbb{P}_0' = \bigvee_{i=0}^{n-1} \phi'^{-i}(\mathbb{P}) \cap F' < \mathbb{Q}_0'.$$

This is clearly possible to do by partitioning each of the atoms of $\mathbb{P}_0'$ separately. Now let $\mathbb{Q}_1'$ be the lumping of atoms of $\mathbb{Q}_0'$ corresponding to the recovery of $\mathbb{Q}$ from $\mathbb{P} \vee \mathbb{Q}$ under the correspondence (3), and define $\mathbb{Q}'$ from $\mathbb{Q}_1'$ as we did $\mathbb{P}'$ from $\mathbb{Q}'$ in the previous proof. ∎

Proof (Lemma 5.6) Note that the partitions $\mathbb{P}$ and $\mathbb{P}'$ are only given on $H = \bigcup_{i=0}^{n-1} \phi^i F$ and $H' = \bigcup_{i=0}^{n-1} \phi'^i F'$, respectively. Thus the right sides of (a) and (b) refer to distributions on those sets. Further note that the conclusion of the lemma is equivalent to asserting the existence of a partition $\mathbb{Q}$ such that Γ' is isomorphic to the gadget $(F, \phi, n, \mathbb{Q})$ and

$$\frac{1}{n}\sum_{i=0}^{n-1} \rho(\mathbb{P} \cap \phi^i F, \mathbb{Q} \cap \phi^i F) < 3\varepsilon. \tag{4}$$

The proof is by induction on n. For $n = 1$ the existence of a $\mathbb{Q}$ satisfying (4), which becomes

$$\rho(\mathbb{P} \cap F, \mathbb{Q} \cap F) = \sum_{i=1}^{l} \mu(P_i \,\Delta\, Q_i) < 3\varepsilon,$$

is immediate from (e):

$$\delta(\mathbb{P}, \mathbb{P}') = \sum_{i=1}^{l} |\mu(P_i) - \mu'(P_i')| < \varepsilon.$$

Indeed, a copy $\mathbb{Q}$ of $\mathbb{P}'$ can be constructed on F such that for each $i = 1, \ldots, l$ either $P_i \subseteq Q_i$ or $Q_i \subseteq P_i$. Then $d(\mathbb{Q}) = d(\mathbb{P}')$ so that $(F, \phi, 1, \mathbb{Q})$ is isomorphic to $\Gamma' = (F', \phi', 1, \mathbb{P}')$, and

$$\mu(P_i \,\Delta\, Q_i) = |\mu(P_i) - \mu(Q_i)| = |\mu(P_i) - \mu(P_i')|.$$

Assume the theorem is true for $n = k$, and let $\Gamma = (F, \phi, k + 1, \mathbb{P})$, $\Gamma' = (F', \phi', k + 1, \mathbb{P}')$ be gadgets satisfying (a)–(e) for $n = k + 1$. Since each of the conditions (a)–(e) implies the same condition for

$$\Gamma_0 = \left(F, \phi, k, \mathbb{P} \cap \left(\bigcup_{i=0}^{k-1} \phi^i F\right)\right) \quad \text{and} \quad \Gamma_0' = \left(F', \phi', k, \mathbb{P}' \cap \left(\bigcup_{i=0}^{k-1} \phi'^i F'\right)\right),$$

it follows that there exists a partition $\mathbb{Q}_0$ on $\bigcup_{i=0}^{k-1} \phi^i F$ such that $(F, \phi, k, \mathbb{Q}_0)$ is isomorphic to Γ_0' and

$$\frac{1}{k} \sum_{i=0}^{k-1} \rho(\mathbb{P} \cap \phi^i F, \mathbb{Q}_0 \cap \phi^i F) < 3\varepsilon. \tag{5}$$

We need to show that $\mathbb{Q}_0$ can be extended to a partition $\mathbb{Q}$ by defining it on the top level $\phi^k F$ in an appropriate manner so that (4) holds for $n = k + 1$ and $\Gamma'' = (F, \phi, k + 1, \mathbb{Q})$ is isomorphic to Γ'.

We begin by extending the columns of the gadgets Γ_0 and $(F, \phi, k, \mathbb{Q}_0)$ to $\phi^k F$. That is, let

$$\mathbb{P}_k = \phi^k\left(\left(\bigvee_{i=0}^{k-1} \phi^{-i}\mathbb{P}\right) \cap F\right) \qquad \text{and} \qquad \mathbb{Q}_k = \phi^k\left(\left(\bigvee_{i=0}^{k-1} \phi^{-i}\mathbb{Q}_0\right) \cap F\right)$$

be the corresponding partitions of $\phi^k F$. In order that Γ'' be isomorphic to Γ' we need to have $\mathbb{Q} \cap \phi^k F$ independent of $\mathbb{Q}_k$ [so that $\phi^{-i}\mathbb{Q} \cap F$ $(i = 0, \ldots, k)$ will be an independent sequence]. This is accomplished by defining $\mathbb{Q}$ on $A \in \mathbb{Q}_n$ such that $d(\mathbb{Q} \cap A) = d(\mathbb{P}' \cap \phi'^k F')$. This will also give the desired distribution for $\mathbb{Q} \cap \phi^k F$. However, it will not guarantee that (4) holds.

In order to obtain the closeness of fit desired between $\mathbb{Q} \cap \phi^k F$ and $\mathbb{P} \cap \phi^k F$, we first replace Γ by an isomorphic gadget Γ''' that differs from Γ only on the top level and satisfies

$$d(\mathbb{P}''' \cap (A \cap B)) = d(\mathbb{P}''' \cap B) \tag{6}$$

for each $A \in \mathbb{Q}_n$, $B \in \mathbb{P}_n$. This we do by defining $\mathbb{P}'''$ on $\phi^k F$ so that it partitions each $A \cap B$ $(A \in \mathbb{Q}_n, B \in \mathbb{P}_n)$ the same as $\mathbb{P}$ partitions B, that is, $d(\mathbb{P}''' \cap (A \cap B)) = d(\mathbb{P} \cap B)$. Let us drop the triple prime and assume $\mathbb{P}$ satisfies (6).

According to (c) and the definition of ε-independence, there exists a set $C \in \bigvee_{i=0}^{k-1} \phi^{-i}\mathbb{P} \cap F$ such that $\mu(C) \geq (1 - \varepsilon)\mu(F)$ and

$$\delta(\phi^{-k}\mathbb{P} \cap A, \phi^{-k}\mathbb{P}) < \varepsilon, \qquad A \in \bigvee_{i=0}^{k-1} \phi^{-i}\mathbb{P} \cap C.$$

It follows that

$$\delta(\mathbb{P} \cap B, \mathbb{P}) < \varepsilon, \qquad B \in \mathbb{P}_k \cap \phi^k C.$$

Combining this with (e) gives

$$\delta(\mathbb{P} \cap (A \cap B), \mathbb{P}') < 2\varepsilon, \qquad A \in \mathbb{Q}_k, \quad B \in \mathbb{P}_k \cap \phi^k C.$$

As before this means we can now construct $\mathbb{Q}$ on $\phi^k F$ so that

(i) $d(\mathbb{Q} \cap (A \cap B)) = d(\mathbb{P}' \cap \phi'^k F')$, $A \in \mathbb{Q}_n$, $B \in \mathbb{P}_n$, and
(ii) $\rho(\mathbb{Q} \cap (A \cap B), \mathbb{P} \cap (A \cap B)) < 2\varepsilon$, $A \in \mathbb{Q}_n$, $B \in \mathbb{P}_n \cap \phi^k C$.

We have thus constructed $\mathbb{Q} \cap \phi^k F$ to have the same distribution as $\mathbb{P}' \cap \phi'^k F'$ and to be independent of $\mathbb{Q}_k$. Thus $(F, \phi, k+1, \mathbb{Q})$ is isomorphic to $(F', \phi', k+1, \mathbb{P}')$. Moreover, by (ii) and $\mu(\phi^k C) = \mu(C) \geq (1-\varepsilon)\mu(F)$ we have that

$$\rho(\mathbb{Q} \cap \phi^k F, \mathbb{P} \cap \phi^k F) < 3\varepsilon,$$

and the induction step is complete. ∎

Proof (Lemma 5.7) This follows from Theorem 4.13 in much the same way as the proof of Lemma 5.1 does from the ergodic theorem. Indeed, since L_1 convergence implies convergence in measure, given $\varepsilon > 0$ we have

$$\lim_{n\to\infty} \mu\left\{x : \left| h(\phi, \mathbb{P}) - \frac{1}{n} I\left(\bigvee_{i=0}^{n-1} \phi^{-i}\mathscr{B}(\mathbb{P}), x\right) \right| \geq \varepsilon\right\} = 0.$$

If D_n is the indicated set with $\mu(D_n) \to 0$, then

$$C_n = X \sim D_n \in \bigvee_{i=0}^{n-1} \phi^{-i}\mathscr{B}(\mathbb{P})$$

is the union of a collection $\mathscr{C}_n$ of atoms of $\bigvee_{i=0}^{n-1} \phi^{-i}\mathbb{P}$. Finally, for $x \in A \in \bigvee_{i=0}^{n-1} \phi^{-i}\mathbb{P}$ we have $I(\bigvee_{i=0}^{n-1} \phi^{-i}\mathscr{B}(\mathbb{P}), x) = -\log A$. ∎

Notice the difference between Lemma 5.7 and Corollary 4.13.1, which implies that for any atom A of $\bigvee_{i=0}^{n-1} \phi^{-i}\mathbb{P}$ and sufficiently large n we have $|h(\phi, \mathbb{P}) + (1/n) \log \mu(A)| < \varepsilon/\mu(A)$.

Proof (Lemma 5.8) We need to show that $\phi^n\mathbb{P}$ is ε-independent of $\bigvee_{i=0}^{n-1} \phi^i\mathbb{P}$, or equivalently that $\mathbb{P}$ is ε-independent of $\bigvee_{i=0}^{n-1} \phi^{-i}\mathbb{P}$, for each positive integer n. Since

$$H(\mathbb{P}) - h(\phi, \mathbb{P}) \geq H(\mathbb{P}) - H\left(\mathbb{P} \,\middle|\, \bigvee_{i=0}^{n-1} \phi^{-i}\mathbb{P}\right),$$

it is sufficient to show that there exists a $\delta > 0$ such that, if $\mathbb{P}$ and $\mathbb{Q}$ are partitions with $\mathbb{P}$ having k atoms and

$$H(\mathbb{P}) - H(\mathbb{P} \mid \mathbb{Q}) \leq \delta,$$

then $\mathbb{P}$ is ε-independent of $\mathbb{Q}$.

Moreover, we may assume that $\mathbb{Q}$ is a two-set partition. For if $\mathbb{P}$ has k sets and is not ε-independent of $\mathbb{Q}$, then the collection $\mathscr{C}$ of atoms A of $\mathbb{Q}$ for which $\delta(\mathbb{P} \cap A, \mathbb{P}) \geq \varepsilon$ has total measure greater than ε. It follows that

$$\sum_{P \in \mathbb{P}} \sum_{A \in \mathscr{C}} |\mu(P \cap A) - \mu(P)\,\mu(A)| \geq \varepsilon^2.$$

Then there must be a $P_0 \in \mathbb{P}$ such that

$$\sum_{A \in \mathscr{C}} |\mu(P_0 \cap A) - \mu(P_0)\,\mu(A)| \geq \frac{\varepsilon^2}{k}.$$

If $\mathscr{C}^+$ and $\mathscr{C}^-$ denote the subcollections of $A \in \mathscr{C}$ for which $\mu(P \cap A) - \mu(P)\,\mu(A)$ is, respectively, nonnegative and negative, it follows that for one of them, call it $\mathscr{C}'$,

$$\left| \sum_{A \in \mathscr{C}'} [\mu(P_0 \cap A) - \mu(P_0)\,\mu(A)] \right| \geq \frac{\varepsilon^2}{2k}.$$

Let $S = \cap\, \mathscr{C}'$ and $\mathbb{Q}_0 = \{S, \tilde{S}\}$. Then

$$\mu(S) \geq \varepsilon^2/2k \qquad \text{and} \qquad \delta(\mathbb{P} \cap S, \mathbb{P}) \geq \varepsilon^2/2k \tag{7}$$

and

$$0 < H(\mathbb{P}) - H(\mathbb{P} \mid \mathbb{Q}_0) \leq H(\mathbb{P}) - H(\mathbb{P} \mid \mathbb{Q}). \tag{8}$$

Thus $\mathbb{P}$ is not $(\varepsilon^2/2k)$-independent of $\mathbb{Q}_0$, and if $H(\mathbb{P}) - H(\mathbb{P} \mid \mathbb{Q}_0) > \delta$, so is $H(\mathbb{P}) - H(\mathbb{P} \mid \mathbb{Q})$.

Now let $K \subseteq R^{3k+1}$ be the set of all vectors $(d(\mathbb{P}),\ d(\mathbb{P} \cap S),\ d(\mathbb{P} \cap \tilde{S}),\ \mu(S))$ determined by k-set partitions $\mathbb{P}$ of X and sets $S \in \mathscr{B}$ for which (7) holds. Clearly, $\mathbb{P}$ cannot be refined by $\mathbb{Q}_S = \{S, \tilde{S}\}$, and so $H(\mathbb{P}) - H(\mathbb{P} \mid \mathbb{Q}_S)$ is never zero on K. Since K is compact and $H(\mathbb{P}) - H(\mathbb{P} \mid \mathbb{Q}_S)$ is continuous on K, it is bounded away from zero, and the proof is complete. ▮

Proof (Lemma 5.9) For a fixed number of atoms the function $H(\mathbb{P}')$ is clearly continuous for the distribution metric $\delta(\mathbb{P}, \mathbb{P}')$ (since $-t \log t$ is a continuous function of t). Thus we may choose δ small enough that (i) implies $|H(\mathbb{P}) - H(\mathbb{P}')| < \delta_1$ for any predetermined $\delta_1 > 0$. This with (ii) implies that

$$|H(\mathbb{P}') - h(\phi', \mathbb{P}')| < \delta + \delta_1.$$

According to Lemma 5.8, if δ and δ_1 are sufficiently small, this implies that $\{\phi'^n \mathbb{P}'\}$ is an ε-independent sequence. Now apply Lemma 5.3 to build gadgets Γ and Γ' satisfying the hypotheses of Lemma 5.6. If $\delta < \varepsilon/6$, it follows from that lemma that $\gamma(\Gamma, \Gamma') < \varepsilon/2$. If Γ and Γ' are made to come sufficiently close to filling X and X', respectively, we can conclude that (iii) holds. ▮

Remark For a nonatomic measure space X it is not hard to see that $h(\phi, \mathbb{P})$ assumes every value between 0 and $h(\Phi)$ as $\mathbb{P}$ ranges over the finite partitions of X. Indeed, for any $\mathbb{P}_1 = \{P_1, \dots, P_l\}$ let $\mathbb{P}_t = \{P_{t1}, \dots, P_{tl}\}$ $(0 \leq t < 1)$ be such that

$$\mu(P_{t1}) = 1 - t(1 - \mu(P_1)) \qquad \text{and} \qquad \mu(P_{tj}) = \frac{1 - \mu(P_{t1})}{1 - \mu(P_1)} \mu(P_j)$$

for $j = 2, \dots, l$. Then $h(\phi, \mathbb{P}_t)$ is a continuous function of t and so assumes all values between $h(\phi, \mathbb{P}_0) = 0$ and $h(\phi, \mathbb{P}_1)$.

Proof (Lemma 5.10) Since

$$h(\Phi) = h(\phi, \mathbb{P}) \leq h(\Phi'),$$

it follows from the preceding remark that given $\varepsilon > 0$ we can choose a partition $\mathbb{Q}_0$ of X' such that

$$0 < h(\phi, \mathbb{P}) - h(\phi', \mathbb{Q}_0) < \varepsilon.$$

Now let $\beta > 0$ and choose n sufficiently large (Lemma 5.7) that

(a) there is a collection $\mathscr{C}' \subseteq \bigvee_{i=0}^{n-1} \phi'^{-i}\mathbb{Q}_0$ with

  (i) $\mu(\cup \mathscr{C}') \geq 1 - \beta$,
  (ii) $e^{-n(h(\phi', \mathbb{Q}_0)+\beta)} \leq \mu(A) \leq e^{-n(h(\phi', \mathbb{Q}_0)-\beta)}$ for $A \in \mathscr{C}'$; and

(b) there is a collection $\mathscr{C} \subseteq \bigvee_{i=0}^{n-1} \phi^{-i}\mathbb{P}$ with

  (i) $\mu(\cup \mathscr{C}) \geq 1 - \beta$,
  (ii) $e^{-n(h(\phi, \mathbb{P})+\beta)} \leq \mu(A) \leq e^{-n(h(\phi, \mathbb{P})-\beta)}$ for $A \in \mathscr{C}$.

By choosing β small enough and n large enough, since $h(\phi', \mathbb{Q}_0) < h(\phi, \mathbb{P})$, we can assume that

$$e^{n(h(\phi', \mathbb{Q}_0)+\beta)} \leq e^{n(h(\phi, \mathbb{P})-\beta)}(1 - \beta).$$

It follows, from (iii) of Lemma 5.7, that

$$v(\mathscr{C}') \leq v(\mathscr{C}). \tag{9}$$

According to Lemma 5.1, we can assume that n and $\mathscr{C}$ satisfy

$$\sum_{P_i \in \mathbb{P}} |f_A(i, n) - \mu(P_i)| < \beta, \qquad A \in \mathscr{C}, \tag{10}$$

where $f_A(i, n)$ is the relative frequency of occurrence of i in the $\mathbb{P}$-n-name of the atom A. It then follows from Lemma 5.3 that there is a gadget $(F', \phi', n, \mathbb{Q}_0)$ in $(X', \mathscr{B}', \mu')$ such that

$$\mu'\left(\bigcup_{i=0}^{n-1} \phi'^i F'\right) > 1 - \beta \tag{11}$$

and

$$\delta\left(\bigvee_{i=0}^{n-1} \phi'^{-i}\mathbb{Q}_0 \cap F', \bigvee_{i=0}^{n-1} \phi'^{-i}\mathbb{Q}_0\right) < 2\beta. \tag{12}$$

Now let $F = (\cup \mathscr{C}') \cap F'$, and set $\mathscr{C}_F' = \mathscr{C}' \cap F$. From (i) of part (a), (9), and (11), it follows that

$$\mu'\left(\bigcup_{i=0}^{n-1} \phi'^{i}F\right) > 1 - 2\beta \tag{13}$$

and

$$\nu(\mathscr{C}) \geq \nu(\mathscr{C}_F'). \tag{14}$$

Let ψ be any monic map of $\mathscr{C}_F'$ into $\mathscr{C}$. We define a new partition $\mathbb{Q}$ on $\bigcup_{i=0}^{n-1} \phi'^{i}F$ so that $(F, \phi', n, \mathbb{Q})$ has the same column structure as $(F, \phi', n, \mathbb{Q}_0)$ as follows. For each $A \in \mathscr{C}_F'$ assign the column level $\phi'^{j}A$ to Q_{i_j} where $\psi(A) = P_{i_1} \cap \phi^{-1}P_{i_2} \cap \cdots \cap \phi^{-(n-1)}P_{i_{n-1}}$. Define $\mathbb{Q}$ arbitrarily on $X \sim \bigcup_{i=0}^{n-1} \phi'^{i}F$ without increasing the number of atoms beyond $\nu(\mathbb{P})$. Clearly, $\mathbb{Q}$ satisfies conclusion (i) of the lemma. Let us show that it also satisfies (ii) and (iii) if α, β and n are chosen appropriately.

If $A \in \mathscr{C}_F'$ and if $\bar{A} = \bigcup_{i=0}^{n-1} \phi^{i}A$ is the column over A, we have by (10) that

$$\sum_{P_i \in \mathbb{P}} |f_{\psi A}(i, n) - \mu(P_i)| < \beta,$$

which implies

$$\sum_{P_i \in \mathbb{P}} |\mu'(Q_i \cap \bar{A}) - \mu(P_i)\, \mu'(A)| < \beta\mu'(A).$$

Together with (13) this gives (ii) for the proper choice of β.

Now let $\mathbb{Q}_1 = \{F, \tilde{F})$. Then, recalling that $(F, \phi', n, \mathbb{Q})$ and $(F, \phi', n, \mathbb{Q}_0)$ have the same column structure, we see easily that

$$\mathbb{Q}_0 \cap \bar{F} < \left(\bigvee_{i=-n}^{n} \phi'^{i}(\mathbb{Q} \vee \mathbb{Q}_1)\right) \cap \bar{F}$$

and

$$\mathbb{Q} \cap \bar{F} < \left(\bigvee_{i=-n}^{n} \phi'^{i}(Q_0 \vee Q_1)\right) \cap \bar{F}.$$

If β is small, $\bar{F}$ is almost all of X'. If also n is large, $\mathbb{Q}_1$ is almost trivial. Hence $h(\phi', \mathbb{Q}_0)$ is near $h(\phi', \mathbb{Q})$. Thus we can conclude from

$$0 < h(\phi, \mathbb{P}) - h(\phi', \mathbb{Q}_0) < \varepsilon$$

that proper choices of n and β give

$$0 < h(\phi, \mathbb{P}) - h(\phi', \mathbb{Q}) < \varepsilon. \blacksquare$$

Proof (Lemma 5.11) Since $\mathbb{P}$ is finitely determined, there exist n_1 and $\delta_1 > 0$ such that (i)–(iv) imply $\pi((\Phi, \mathbb{P}), (\Phi', \mathbb{P}')) < \varepsilon_1$ (to be determined later). We may assume that $\delta_1 < \varepsilon$. Furthermore, we can assume that

$$0 < h(\phi, \mathbb{P}) - h(\phi', \mathbb{P}') < \delta_1. \tag{15}$$

Otherwise, it would be true that $h(\phi, \mathbb{Q}) = h(\phi', \mathbb{P}')$ for all $\mathbb{P}'$ satisfying (ii) and (iii). In this case, we could choose $\mathbb{Q}$ satisfying (ii), (vii), and $d(\bigvee_{i=0}^{n-1} \phi'^{-i}\mathbb{Q}) = d(\bigvee_{i=0}^{n-1} \phi^{-i}\mathbb{P})$, and be done.

Now choose as in the proof of Lemma 5.10 a partition $\mathbb{Q}_0$ such that $\mathbb{P}' < \mathbb{Q}_0$ and $0 < h(\phi, \mathbb{P}) - h(\phi', \mathbb{Q}_0) < \alpha$, where α is yet to be specified. Next choose $\beta > 0$ and n such that (a), (b), (9), and (10) all hold, and choose F' so that (11) and (12) hold. It follows then that

$$\delta\left(\bigvee_{i=0}^{n-1} \phi'^{-i}\mathbb{P}' \cap F', \bigvee_{i=0}^{n-1} \phi'^{-i}\mathbb{P}'\right) < 2\beta.$$

Define F and $\mathscr{C}_F'$ as before, so that (13) and (14) hold. Now define $\psi: \mathscr{C}_F' \to \mathscr{C}$ as before but subject to the following condition:

(A) for some collection $\mathscr{A} \subseteq \mathscr{C}_F'$ with $\mu'(\cup \mathscr{A}) \geq (1 - \varepsilon/3)\,\mu'(F)$ it is true that the $\mathbb{P}'$-n-name of each $A \in \mathscr{A}$ and the $\mathbb{P}$-n-name of $\psi(A) \in \mathscr{C}$ agree in more than $n\varepsilon/3$ places.

It will then follow that

$$\sum_i \mu[(Q_i \cap \overline{A}) \,\Delta\, (P_i' \cap \overline{A})] \leq \frac{\varepsilon}{3}\,\mu(\overline{A}),$$

where $\overline{A}$ is the column over $A \in \mathscr{C}$, so that

$$\rho(\mathbb{Q} \cap \overline{F}, \mathbb{P}' \cap \overline{F}) \leq 2\varepsilon/3.$$

Thus to complete the proof of the lemma we only need to show that ψ can be defined in such a way that (A) holds.

To establish (A) we go back to the inequality

$$\pi((\Phi, \mathbb{P}), (\Phi', \mathbb{P}')) < \varepsilon_1,$$

with its still unspecified ε_1. Choose a gadget (F, ϕ, n, P) in X such that

$$d\left(\bigvee_{i=0}^{n-1} \phi^{-i}\mathbb{P} \cap F_0\right) = d\left(\bigvee_{i=0}^{n-1} \phi^{-i}\mathbb{P}\right).$$

By Lemma 5.3 choose a new partition $\mathbb{P}^*$ so that $(F', \phi', n, \mathbb{P}^*)$ is isomorphic to $(F_0, \phi, n, \mathbb{P})$ and

$$\frac{1}{n}\sum_{i=0}^{n-1} \rho(\mathbb{P}' \cap \phi'^{i}F', \mathbb{P}^* \cap \phi'^{i}F') < \varepsilon_1.$$

Replace F' by F. For sufficiently small β

$$\delta\left(\bigvee_{i=0}^{n-1} \phi^{-i}\mathbb{P} \cap F_0, \bigvee_{i=0}^{n-1} \phi'^{-i}\mathbb{P}^* \cap F\right) < \beta$$

and

$$\frac{1}{n}\sum_{i=0}^{n-1} \rho(\mathbb{P}' \cap \phi'^i F, \mathbb{P}^* \cap \phi'^i F) < 2\varepsilon_1.$$

Put $\mathscr{C}_F{}^* = \mathscr{C}^* \cap F$. For small β we can assume $\nu(\mathscr{C}_F{}^*) \geq \nu(\mathscr{C}_F{}')$. In fact, we shall choose β so small that

$$\mu'(A) \geq 4\mu(A^*), \qquad A \in \mathscr{C}_F{}', \qquad A^* \in \mathscr{C}_F{}^*, \tag{16}$$

so that at least 4 sets in $\mathscr{C}_F{}^*$ are needed to cover a set in $\mathscr{C}_F{}'$.

Let $\mathscr{A}$ consist of all those $A \in \mathscr{C}_F{}'$ such that more than half of A is covered by sets $A^* \in \mathscr{C}_F{}^*$ such that the $\mathbb{P}'$-n-name of A and the $\mathbb{P}^*$-n-name of A^* differ in no more than $n\sqrt{2\varepsilon_1}$ places. Let E be the set of points $x \in F$ such that the $\mathbb{P}'$-n-name and $\mathbb{P}^*$-n-name of x disagree in more than $n\sqrt{2\varepsilon_1}$ places. It follows easily that $\mu'(E) \leq \sqrt{2\varepsilon_1}\,\mu'(F)$. Moreover, if $B = \cup\, \mathscr{C}_F{}^*$, then

$$\sum_{A \in \mathscr{C}_F{}' \sim \mathscr{A}} \mu'(A) \leq 2[\mu'(F \sim B) + \mu'(E)].$$

If ε_1 and β are small enough, it follows that $\mu'(\cup\, \mathscr{A}) \geq (1 - \varepsilon/3)\mu'(F)$ as asserted in (A).

It is clear from (16) that any k elements $A_1, \ldots, A_k \in \mathscr{A}$ intersect at least k elements in $\mathscr{C}_F{}^*$ whose $\mathbb{P}^*$-n-name differs from the $\mathbb{P}'$-n-name of at least one of the A_i in no more than $n\sqrt{2\varepsilon_1}$ places. The marriage lemma of combinatorics (see [30]) then implies the existence of a monic map ψ from $\mathscr{A}$ to $\mathscr{C}_F{}^*$ such that the $\mathbb{P}'$-n-name of $A \in \mathscr{A}$ and the $\mathbb{P}^*$-n-name of $\psi(A)$ differ in no more than $n\sqrt{2\varepsilon_1}$ places. Extend the definition of ψ to $\mathscr{C}_F{}^* \sim \mathscr{A}$ in any way, and we have proved (A). ∎

Proof (Lemma 5.12) The δ we choose is the δ of Lemma 5.11 corresponding to $\varepsilon/2$. Let Φ' and $\mathbb{P}'$ be as specified in Lemma 5.12, and for each n let δ_n be the δ of Lemma 5.11 corresponding to $\varepsilon/2^n$. By induction choose partitions $\mathbb{Q}_n$ such that $\mathbb{Q}_0 = \mathbb{P}'$ and (by Lemma 5.11)

(a) $\delta(\mathbb{Q}_n, \mathbb{P}) < \delta_{n+1}$,
(b) $0 \leq h(\phi, \mathbb{P}) - h(\phi', \mathbb{Q}_n) \leq \delta_{n+1}$,
(c) $\rho(\mathbb{Q}_n, \mathbb{Q}_{n-1}) < \varepsilon/2^n$.

Thus $\{\mathbb{Q}_n\}$ is a Cauchy sequence of partitions in the ρ-metric. It follows readily from the Riesz–Fischer theorem that this implies the existence of a limiting partition $\mathbb{Q}$, that is, $\rho(\mathbb{Q}_n, \mathbb{Q}) \to 0$. But then from (a)–(c) we deduce

(d) $d(\mathbb{Q}) = d(\mathbb{P})$,
(e) $h(\phi', \mathbb{Q}) = h(\phi, \mathbb{P})$,
(f) $\rho(\mathbb{Q}, \mathbb{P}') \leq \rho(\mathbb{P}', \mathbb{Q}_0) + \sum_{n=1}^{\infty} \rho(\mathbb{Q}_n, \mathbb{Q}_{n-1}) < \varepsilon$.

Since (d) and (e) imply that

$$h(\phi', \mathbb{Q}) = H(\mathbb{P}) = H(\mathbb{Q}),$$

it follows (Exercise 2) that $\{\phi^n \mathbb{Q}\}$ is an independent sequence, and we are done. ▮

3. THE ISOMORPHISM THEOREM

Theorem 5.1 *(Ornstein) Let $\pi = (p_1, p_2, \ldots, p_k)$ and $\pi' = (p_1', p_2', \ldots, p_l')$ be probability distributions. Then the Bernoulli systems $\Phi(\pi)$ and $\Phi(\pi')$ are isomorphic iff $h(\Phi(\pi)) = h(\Phi(\pi'))$.*

According to Definition 5.1″ and the Remark following, it is sufficient to show that for two systems Φ and Φ' with independent (finite) generators $\mathbb{P}$ and $\mathbb{P}'$, respectively, such that $h(\Phi) = h(\Phi')$, we can find a partition $\mathbb{Q}$ of X such that

(a) $\nu(\mathbb{Q}) = \nu(\mathbb{P}')$,
(b) $d(\mathbb{Q}) = d(\mathbb{P}')$,
(c) $\phi^n \mathbb{Q}$ $(n = 0, 1, 2, \ldots)$ is an independent sequence and
(d) $\mathbb{P} \subseteq \bigvee_{n=-\infty}^{\infty} \mathscr{B}(\phi^n \mathbb{Q})$.

Throughout this section Φ, Φ', $\mathbb{P}$, and $\mathbb{P}'$ will be assumed to be as above.

Lemma 5.12 provides the tool for satisfying (a)–(c) and an approximate form of (d). In order to facilitate the use of this lemma in a succession of approximations, we introduce the following notation.

Definition 5.17 If $\mathbb{P}$ and $\mathbb{Q}$ are partitions of the same space $(X, \mathscr{B}, \mu)$ and $\varepsilon > 0$, we say that $\mathbb{P}$ *ε-refines* $\mathbb{Q}$ and write $\mathbb{Q} <_{\varepsilon} \mathbb{P}$ if there is a partition $\mathbb{Q}' < \mathbb{P}$ such that (a) $\nu(\mathbb{Q}') = \nu(\mathbb{Q})$, and (b) $\rho(\mathbb{Q}, \mathbb{Q}') < \varepsilon$.

Now it is easily seen that condition (d) above is equivalent to the following.

(e) For each $\varepsilon > 0$ there exists $n = n(\varepsilon)$ such that $\mathbb{P} <_{\varepsilon} \bigvee_{i=-n}^{n} \phi^i \mathbb{Q}$.

For a measurable partition $\mathbb{Q}$ of X, let $\mathbb{Q}^{\infty} = \bigvee_{n=-\infty}^{\infty} \mathscr{B}(\phi^n \mathbb{Q})$. Then $\Phi_{\mathbb{Q}} = (X, \mathbb{Q}^{\infty}, \mu, \phi)$ is a factor of Φ ($\Phi_{\mathbb{Q}}$ is the natural extension of $\Phi_{\mathscr{B}(\mathbb{Q})}$ as defined in Chapter IV). Conditions (a)–(c) are sufficient to show that $\Phi_{\mathbb{Q}} \cong \Phi'$. In general, if $\mathbb{Q}$ and $\mathbb{Q}'$ are partitions of X and X', respectively, then $\Phi_{\mathbb{Q}} \cong \Phi'_{\mathbb{Q}'}$ under the canonical correspondence determined by identifying $\mathbb{Q}$ and $\mathbb{Q}'$ iff

(f) $d(\bigvee_{i=0}^{n} \phi^i\mathbb{Q}) = d(\bigvee_{i=0}^{n} \phi'^i\mathbb{Q}')$, $n = 0, 1, 2, \ldots$.

In this case, we shall write $\Phi_{\mathbb{Q}} \sim \Phi'_{\mathbb{Q}'}$. Our next result shows that if $\Phi_{\mathbb{Q}} \sim \Phi'$ we can find a partition $\mathbb{P}_1$ of X such that $\Phi_{\mathbb{P}_1} \sim \Phi_{\mathbb{P}}$, $\Phi_{\mathbb{P}_1}$ is a (canonical) factor of $\Phi_{\mathbb{Q}}$, and $\Phi_{\mathbb{Q}}$ is almost a factor of $\Phi_{\mathbb{P}_1}$. (Note that $\Phi_{\mathbb{P}_1} \sim \Phi_{\mathbb{P}}$ does *not* imply $\mathbb{P}_1$ is a generator for Φ, but only that Φ contains a "copy" of itself.)

Lemma 5.13 *Suppose $\Phi_{\mathbb{Q}} \sim \Phi'_{\mathbb{P}'} = \Phi'$ and let $\varepsilon > 0$. Choose N so that $\mathbb{Q} <_\varepsilon \bigvee_{i=-N}^{N} \phi^i\mathbb{P}$. Then there is a partition $\mathbb{P}_1$ of X such that*

(i) $\Phi_{\mathbb{P}_1} \sim \Phi_{\mathbb{P}}$,
(ii) $\mathbb{P}_1 \subseteq \mathbb{Q}^\infty$, and
(iii) $\mathbb{Q} <_{2\varepsilon} \bigvee_{i=-N}^{N} \phi^i\mathbb{P}_1$.

More precisely, if $\mathbb{Q}' < \bigvee_{i=-N}^{N} \phi^i\mathbb{P}$ *satisfies* $\rho(\mathbb{Q}, \mathbb{Q}') < \varepsilon$, *then the canonical copy* $\mathbb{Q}_1' < \bigvee_{i=-N}^{N} \phi^i\mathbb{P}_1$ *satisfies* $\rho(\mathbb{Q}, \mathbb{Q}_1') < 2\varepsilon$.

Proof Suppose $\mathbb{Q}'$ is as indicated. That is, $\mathbb{Q}' = f(\mathbb{P}) = (f_1(\mathbb{P}), f_2(\mathbb{P}), \ldots, f_k(\mathbb{P}))$ is a partition of X into k sets, each of which is a union of atoms from $\bigvee_{i=-N}^{N} \phi^i\mathbb{P}$, and $\rho(\mathbb{Q}, \mathbb{Q}') < \varepsilon$. Then, for any other partition $\overline{\mathbb{P}}$ of X with $v(\overline{\mathbb{P}}) = v(\mathbb{P})$, $f(\overline{\mathbb{P}})$ makes sense and is a partition of X into k sets with $f(\overline{\mathbb{P}}) < \bigvee_{i=-N}^{N} \phi^i\overline{\mathbb{P}}$.

Now if $\alpha > 0$ is any positive number, there is an $N_1 \geq N$ such that $\mathbb{Q} <_\alpha \bigvee_{i=-N_1}^{N_1} \phi^i\mathbb{P}$. As before, we have $f_1(\mathbb{P}) < \bigvee_{i=-N_1}^{N_1} \phi^i\mathbb{P}$ and $\rho(\mathbb{Q}, f_1(\mathbb{P})) < \alpha$. Applying Lemma 5.2 to $\Phi_{\mathbb{Q}}$, we have for each positive integer n a set $F \in Q^\infty$, depending on α and n, such that $F, \phi F, \ldots, \phi^{n-1}F$ is a disjoint sequence and $\mu(\bigcup_{i=0}^{n-1} \phi^i F) > 1 - \alpha$.

According to Lemma 5.5 applied to $\Phi = \Phi_{\mathbb{P}}$ and $\Phi_{\mathbb{Q}}$, there is a $\mathbb{P}^* \subseteq \mathbb{Q}^\infty$ such that the gadgets $(F, \phi, n, \mathbb{Q} \vee \mathbb{P})$ and $(F, \phi, n, \mathbb{Q} \vee \mathbb{P}^*)$ are isomorphic. This means, in particular, that $\mathbb{Q} \vee \mathbb{P}$ and $\mathbb{Q} \vee \mathbb{P}^*$ have the same distribution on $\overline{F} = \bigcup_{i=0}^{n-1} \phi^i F$. Summing out the atoms of $\mathbb{Q}$, we see that $d(\mathbb{P} \cap \overline{F}) = d(\mathbb{P}^* \cap \overline{F})$. Hence $\delta(P, P^*) < \alpha$.

Let us look at $f_1(\mathbb{P}^*)$. Since $f_1(\mathbb{P}) < \bigvee_{i=-N_1}^{N_1} \phi^i\mathbb{P}$ and $f_1(\mathbb{P}^*) < \bigvee_{i=-N_1}^{N_1} \phi^i\mathbb{P}^*$ are formed by the same rule, it follows that $f_1(\mathbb{P})$ and $f_1(\mathbb{P}^*)$ partition the atoms of $\mathbb{Q}$ in the same way on all of $\overline{F}$ except possibly the top N_1 levels and the bottom N_1 levels. Since $v(\mathbb{Q}) = v(f_1(\mathbb{P})) = v(f_1(\mathbb{P}^*))$, and $\rho(\mathbb{Q}, f_1(\mathbb{P})) < \alpha$, it follows that

$$\rho(\mathbb{Q}, f_1(\mathbb{P}^*)) < \alpha + 2N_1\mu(F) + \alpha < 3\alpha \tag{17}$$

if $n > 2N_1/\alpha$.

Now suppose $\beta > 0$ is given. Then for α sufficiently small, (17) implies that

$$\begin{aligned} h(\phi, \mathbb{P}^*) &\geq h(\phi, f_1(\mathbb{P}^*)) \geq h(\phi, \mathbb{Q}) - \beta \\ &= H(\mathbb{Q}) - \beta = H(\mathbb{P}') - \beta = h(\Phi') - \beta \\ &= h(\Phi) - \beta = h(\phi, \mathbb{P}) - \beta. \end{aligned}$$

If, in addition, $\alpha < \beta$, then we have

$$\delta(\mathbb{P}, \mathbb{P}^*) < \beta, \qquad 0 \leq h(\phi, \mathbb{P}) - h(\phi, \mathbb{P}^*) < \beta. \tag{18}$$

According to Lemma 5.12 (with $\Phi' = \Phi_{\mathbb{Q}}$), if $\delta > 0$ is given, we can choose β so that (18) implies the existence of a partition $\mathbb{P}_1 \subseteq \mathbb{Q}^\infty$ such that $\Phi_{\mathbb{P}_1} \sim \Phi_{\mathbb{P}}$ and $\rho(P_1, P^*) < \delta$.

Finally, given $\varepsilon > 0$ we can choose δ (and then β, α, N, N_1, n in that order) so that $\rho(\mathbb{P}_1, \mathbb{P}^*) < \delta$ implies

$$\rho(\mathbb{Q}, f(\mathbb{P}_1)) \leq \rho(\mathbb{Q}, f(\mathbb{P}^*)) + \rho(f(\mathbb{P}^*), f(\mathbb{P}_1)) < 2\varepsilon,$$

and hence $\mathbb{Q} <_{2\varepsilon} \bigvee_{i=-N}^{N} \phi^i \mathbb{P}_1$. ∎

Lemma 5.14 *Suppose $\Phi_{\mathbb{Q}} \sim \Phi'_{\mathbb{P}'}$, and let $\varepsilon > 0$. Then there is a partition $\mathbb{Q}_1$ of X and a positive integer k such that*

(i) $\Phi_{\mathbb{Q}_1} \sim \Phi'_{\mathbb{P}'}$,
(ii) $\mathbb{P} <_\varepsilon \bigvee_{i=-k}^{k} \phi^i \mathbb{Q}_1$, *and*
(iii) $\rho(\mathbb{Q}, \mathbb{Q}_1) < \varepsilon$.

Proof Given $\alpha > 0$ choose N_1 so that

$$\mathbb{Q} <_\alpha \bigvee_{i=-N_1}^{N_1} \phi^i \mathbb{P}.$$

Apply Lemma 5.13 to find a partition $\mathbb{P}_1$ of X such that $\rho(\mathbb{Q}, f_1(\mathbb{P}_1)) < 2\alpha$, where $f_1(\mathbb{P}_1) < \bigvee_{i=-N_1}^{N_1} \phi^i \mathbb{P}_1$ and $\rho(\mathbb{Q}, f_1(\mathbb{P})) < \alpha$, and such that (i) and (ii) of that lemma hold. Since $\mathbb{P}_1 \subseteq \mathbb{Q}^\infty$, we can choose $N_2 > N_1$ so that

$$\mathbb{P}_1 <_\alpha \bigvee_{i=-N_2}^{N_2} \phi^i \mathbb{Q}.$$

Let f_2 be such that $f_2(\mathbb{Q}) < \bigvee_{i=-N_2}^{N_2} \phi^i \mathbb{Q}$ and $\rho(\mathbb{P}_1, f_2(\mathbb{Q})) < \alpha$.

Next choose $N_3 > N_2$ so that

$$\mathbb{P}_1 <_\beta \bigvee_{i=-N_3}^{N_3} \phi^i \mathbb{Q},$$

where $\beta = \beta(\alpha)$ is yet to be specified. Again let f_3 be such that $f_3(\mathbb{Q}) < \bigvee_{i=-N_3}^{N_3} \phi^i \mathbb{Q}$ and $\rho(\mathbb{P}_1, f_3(\mathbb{Q})) < \beta$.

Given n and $r > 0$ we can by Lemma 5.3 choose $F \in \mathscr{B}$ such that $F, \phi F, \ldots, \phi^{n-1}F$ is a disjoint sequence, $\mu(X \sim \bigcup_{i=0}^{n-1} \phi^i F) < r$, and $\delta((\bigvee_{i=0}^{n-1} \phi^{-i}\mathbb{P}) \cap F, \bigvee_{i=0}^{n-1} \phi^{-i}\mathbb{P}) < 2r$. Similarly, choose $E \in \mathbb{Q}^\infty$ such that $E, \phi E, \ldots, \phi^{n-1}E$ is a disjoint sequence, $\mu(X \sim \bigcup_{i=0}^{n-1} \phi^i E) < r$, and $\delta((\bigvee_{i=0}^{n-1} \phi^{-i}\mathbb{P}_1) \cap E, \bigvee_{i=0}^{n-1} \phi^{-i}\mathbb{P}_1) < 2r$. It follows (since $\Phi_{\mathbb{P}_1} \sim \Phi_{\mathbb{P}}$) that the gadgets $(F, \phi, n, \mathbb{P})$ and $(F, \phi, n, \mathbb{P}_1)$ are isomorphic. Again applying Lemma 5.5, we can find a partition $\mathbb{Q}^* \subseteq \mathscr{B}$ such that $(F, \phi, n, \mathbb{P} \vee \mathbb{Q}^*) \cong (E, \phi, n, \mathbb{P}_1 \vee \mathbb{Q})$.

In the applications of Lemma 5.3 above, let us assume that n is so large and r so small that $\rho(f_3(\mathbb{Q}^*), \mathbb{P}) < 2\beta$ and $\delta(\mathbb{Q}^*, \mathbb{Q}) < 2\beta$. Likewise, we can assume that $\rho(f_2(\mathbb{Q}^*), \mathbb{P}) < 2\alpha$ and $\rho(f_1(\mathbb{P}), \mathbb{Q}^*) < 3\alpha$. The first two conditions imply for β sufficiently small that $h(\phi, \mathbb{Q}^*)$ and $h(\phi, \mathbb{Q})$ are close, and so by Lemma 5.12 that there exists a partition $\mathbb{Q}_1$ satisfying $\Phi_{\mathbb{Q}_1} \sim \Phi_{\mathbb{Q}}$ and $\rho(\mathbb{Q}_1, \mathbb{Q}^*) < \delta$, where $\delta > 0$ is yet to be specified. It will then follow that $\rho(\bigvee_{i=-N_2}^{N_2} \phi^i\mathbb{Q}_1, \bigvee_{i=-N_2}^{N_2} \phi^i\mathbb{Q}^*)$ is small enough (for appropriate choice of δ) that $\rho(f_2(\mathbb{Q}_1), \mathbb{P}) < 3\alpha$.

Summing up, we can choose β such that

$$\begin{aligned} \rho(\mathbb{Q}, \mathbb{Q}_1) &\le \rho(\mathbb{Q}, f_1(\mathbb{P})) + \rho(f_1(\mathbb{P}), \mathbb{Q}^*) + \rho(\mathbb{Q}^*, \mathbb{Q}_1) \\ &< \alpha + 3\alpha + \delta < 5\alpha. \end{aligned}$$

Setting $\alpha = \varepsilon/5$ and $k = N_2$ completes the proof. ▮

Proof of Theorem 5.1 As in the proof of the principal lemma, we make a sequence of approximations and let $\mathbb{Q}$ be the limiting partition. Thus from Lemmas 5.10 and 5.12 we can choose $\mathbb{Q}_0$ so that $\Phi_{\mathbb{Q}_0} \sim \Phi'_{\mathbb{P}'} = \Phi'$. By Lemma 5.14 we can define inductively partitions $\mathbb{Q}_n$ and integers $k_n \uparrow \infty$, such that

(α) $\Phi_{\mathbb{Q}_n} \sim \Phi'_{\mathbb{P}'}$.
(β) $\mathbb{P} <_{2^{-n}} \bigvee_{i=-k_n}^{k_n} \phi^i\mathbb{Q}_n$, and
(γ) $\rho(\mathbb{Q}_{n-1}, \mathbb{Q}_n) < 2^{-n}$.

Let $\mathbb{Q}$ be the limiting partition, $\rho(\mathbb{Q}, \mathbb{Q}_n) \to 0$. It follows from ($\alpha$) that $\Phi_{\mathbb{Q}} \sim \Phi'_{\mathbb{P}'}$, so that (a), (b), and (c) following the statement of the theorem hold. It only remains to show that (d) holds, that is,

$$\mathbb{P} \subseteq \bigvee_{n=-\infty}^{\infty} \mathscr{B}(\phi^n\mathbb{Q}). \tag{19}$$

From (β) and (γ) we have

$$\rho(\mathbb{Q}_n, \mathbb{Q}) \le \sum_{k=n+1}^{\infty} 2^{-k} = 2^{-n}$$

and so

$$\mathbb{P} <_{2^{-n}(1+\nu(\mathbb{P}'))} \bigvee_{i=-k_n}^{k_n} \phi^i \mathbb{Q}. \tag{20}$$

Since (20) implies (e), which is equivalent to (d), we are finished. ▮

4. EXTENSIONS AND CONSEQUENCES OF THE ISOMORPHISM THEOREM

In Section 1 we defined the Bernoulli system $\Phi(\pi)$ where $\pi = (p_1, p_2, p_3, \ldots)$ is a countably infinite probability distribution on Z^+. In this case also,

$$h(\Phi(\pi)) = -\sum_{i=1}^{\infty} p_i \log p_i. \tag{21}$$

Alternatively, we can consider a dynamical system Φ with a countable partition $\mathbb{P}$, such that $\bigvee_{n=-\infty}^{\infty} \mathscr{B}(\phi^n \mathbb{P}) = \mathscr{B}$ and the sequence $\phi^n \mathbb{P}$ is independent.

It can be shown [59] that when the series (21) converges to a finite value, Φ has a *finite* independent generator, so that this case is reduced to that of Theorem 5.1. On the other hand, Ornstein has shown [46] that all Bernoulli systems with infinite entropy are isomorphic.

The following results are due to Ornstein and will not be proved here.

Theorem 5.2 *Every factor of a Bernoulli system is a Bernoulli system.*

Theorem 5.3 *If* $\Phi = \operatorname{inv} \lim_{n\to\infty} \Phi_n$, *and if each* Φ_n *is Bernoulli, then* Φ *is Bernoulli.*

It is, of course, completely obvious that a product of a finite number of Bernoulli systems and hence, by Theorem 5.3, of a countable number of Bernoulli systems is Bernoulli. Thus the class of Bernoulli systems is closed under most of the constructions of Chapter I.

It might be thought that the system Φ_A induced by an ergodic dynamical system Φ is Bernoulli iff Φ is Bernoulli. However, L. Swanson has shown in her Ph.D. dissertation (U. Calif., Berkeley 1975) that certain non-Bernoulli Kolmogorov systems induce Bernoulli systems Φ_A on sets A of measure arbitrarily near one. Moreover, it is known [20] that Φ may be Bernoulli and Φ_A not.

On the positive side, Saleski has shown [54] that a Bernoulli system Φ induces Bernoulli systems Φ_A on uncountably many measure-theoretically distinct sets A, and that, for weakly mixing systems Φ and certain sets A, Φ is Bernoulli if Φ_A is.

For years it was not known if there existed Kolmogorov systems that were not isomorphic to Bernoulli systems, or if entropy was a complete invariant for Kolmogorov systems, which if true would imply that all Kolmogorov systems are Bernoulli. (Why?) However, Ornstein, using techniques developed to prove Theorem 5.1, has shown [48] that there are such Kolmogorov systems. By refining Ornstein's construction, J. Clark showed (unpublished) that there is a Kolmogorov system Φ that has no nth root for any positive integer n. On the other hand, it follows from Theorem 5.1 (see Exercise 3) that Bernoulli systems Φ have roots of all orders. In fact [47], Φ can be embedded in a flow Φ_t such that each Φ_t is Bernoulli and $\Phi_1 = \Phi$.

The remainder of this section is devoted to a brief description of some additional concepts that have proved useful in the identification of Bernoulli systems.

In Section 2, Definition 5.16, we introduced the notion of finitely determined partitions. It was noted that independent generators are finitely determined. We have defined Bernoulli systems to be systems with independent generators. Thus the class of systems with finitely determined generators might be thought to be a larger class. However, it can be shown that they coincide. In fact (see [57]), (i) if Φ has a finitely determined generator, then Φ is Bernoulli, and (ii) if Φ is Bernoulli, *every* generator for Φ is finitely determined.

Statement (i) above is established by carrying out the proof of Theorem 5.1 for systems with finitely determined generators and noting that for every positive number t there is a Bernoulli system Φ with $h(\Phi) = t$.

In [57] it is asserted that *any* partition (not only generators) for a Bernoulli system is finitely determined. Coupled with a theorem of Krieger that systems with finite entropy have finite generators, this and statement (i) above provide a quick proof of Theorem 5.2.

The principal usefulness of the concept of finitely determined partitions comes from the fact that certain conditions on generators, weaker than independence, can be shown to imply the property of being finitely determined. A couple of these, originally contrived by Friedman and Ornstein in the study of Markov systems (Exercise IV.12) and Bernoulli flows are introduced below. The knowledgeable reader will note also the relation to channels with finite memory in information theory (see, e.g., Feinstein [19]).

Definition 5.18 Let $\Phi = (X, \mathscr{B}, \mu, \phi)$ be an ergodic dynamical system. A measurable partition $\mathbb{P}$ of X is said to be *weak Bernoulli* if for each

$\varepsilon > 0$ there is a positive integer N such that $\bigvee_{i=-m}^{0} \phi^i \mathbb{P}$ is ε-independent of $\bigvee_{i=N}^{N+m} \phi^i \mathbb{P}$ for each $m = 0, 1, 2, \ldots$.

Remark $\mathbb{P}$ is Bernoulli (or independent) if we can take $N = 1$ for each ε. Thus Bernoulli partitions are weak Bernoulli. Moreover, the Borel zero–one law shows that weak Bernoulli lies somewhere between Bernoulli and Kolmogorov.

Example 1 If Φ is the Markov system defined in Exercise IV.12, if Φ is mixing so that $\lim_{n\to\infty} p_{ij}^{(n)}$ exists for each i and j [$P^n = (p_{ij}^{(n)})$ is the nth power of P], and if $\mathbb{P} = \{P_1, \ldots, P_l\}$, where $P_j = \{x \in X : x_0 = j\}$, then $\mathbb{P}$ is weak Bernoulli.

In order to give the next definition, let us first expand the domain of the gadget metric γ to arbitrary pairs of finite sequences of partitions. Thus we set

$$\gamma((\mathbb{P}_0, \ldots, \mathbb{P}_{n-1}), (\mathbb{Q}_0, \ldots, \mathbb{Q}_{n-1})) = \inf \frac{1}{n} \sum_{i=0}^{n-1} \rho(\mathbb{P}_i', \mathbb{Q}_i'), \tag{22}$$

where the infimum is over measurable partitions $\mathbb{P}_0', \ldots, \mathbb{P}_{n-1}', \mathbb{Q}_0', \ldots, \mathbb{Q}_{n-1}'$ of $[0, 1]$ satisfying

$$d\left(\bigvee_{i=0}^{n-1} \mathbb{P}_i'\right) = d\left(\bigvee_{i=0}^{n-1} \mathbb{P}_i\right), \qquad d\left(\bigvee_{i=0}^{n-1} \mathbb{Q}_i'\right) = d\left(\bigvee_{i=0}^{n-1} \mathbb{Q}_i\right). \tag{23}$$

Definition 5.19 Let $\Phi = (X, \mathscr{B}, \mu, \phi)$ be an ergodic dynamical system. A measurable partition $\mathbb{P}$ of X is *very weak Bernoulli* if for each $\varepsilon > 0$ there is a positive integer $N = N(\varepsilon)$ such that for each positive integer m there is a collection $\mathscr{C}_m$ of atoms in $\bigvee_{i=-m}^{-1} \phi^i \mathbb{P}$ with

$$\gamma((\mathbb{P} \cap A, \phi\mathbb{P} \cap A, \ldots, \phi^{n-1}\mathbb{P} \cap A), (\mathbb{P}, \phi\mathbb{P}, \ldots, \phi^{n-1}\mathbb{P})) < \varepsilon \tag{24}$$

for each $A \in \mathscr{C}_m$, and

$$\mu(\cup \mathscr{C}_m) > 1 - \varepsilon. \tag{25}$$

It can be shown that weak Bernoulli implies very weak Bernoulli (Exercise 5) and the latter implies finitely determined (see [57]). This is done by generalizing and extending the proof of Lemma 5.9.

EXERCISES

1. Let $\Phi = (X, \mathscr{B}, \mu, \phi)$ be an invertible dynamical system. For each $n = 1, 2, \ldots$ the set $A_n = \{x \in X : \phi^n x = x\}$ is ϕ-invariant. If Φ is ergodic, conclude that $\mu(A_n) = 0$ for all n and Φ is antiperiodic or $\mu(A_n) = 1$ for some

n. In the latter case, show that X is atomic with n atoms, and hence that $h(\Phi) = 0$.

2. If $\{\phi^n \mathbb{P}\}$ is an independent sequence, it follows easily that $h(\phi, \mathbb{P}) = H(\mathbb{P})$. Show that the converse is true.

3. **(a)** If Φ is a Bernoulli system, show that $\Phi^n = (X, \mathscr{B}, \mu, \phi^n)$ is Bernoulli for each n.

(b) Deduce from Exercise IV.13(b) and Theorem 5.1 that Bernoulli systems have roots of all orders, and that the roots are also Bernoulli; that is, given a positive integer n and a Bernoulli system Ψ, there exists a Bernoulli system Φ such that $\Phi^n = \Psi$.

4. Verify the statement in Example 5.1.

5. Let us say that the sequence $\{\mathbb{P}_n\}$ of partitions of X is *weakly ε-independent* if for each $n \in Z^+$ there is a collection $\mathscr{C}_n \subseteq \mathbb{P}_n$ such that

(i) $\mu(\cup \mathscr{C}_n) > 1 - \varepsilon$, and

(ii) $\gamma((\mathbb{P}_0 \cap A, \mathbb{P}_1 \cap A, \ldots, \mathbb{P}_{n-1} \cap A), (\mathbb{P}_0, \mathbb{P}_1, \ldots, \mathbb{P}_{n-1})) < \varepsilon$ for all $A \in \mathscr{C}_n$.

(a) Show that an ε-independent sequence is 2ε-weakly independent.

(b) Show that a weak Bernoulli partition is very weak Bernoulli.

Bibliography

CITED REFERENCES

1. L. M. Abramov, Entropy of a derived automorphism, *Dokl. Akad. Nauk SSSR* **128** (1959), 647–650 [*Amer. Math. Soc. Transl.* Ser. II, **49** (1960), 162–176].
2. L. M. Abramov, Metric automorphisms with quasidiscrete spectrum, *Izv. Akad. Nauk SSSR Ser. Mat.* **26** (1962), 513–530 [*Amer. Math. Soc. Transl.* (2) **39** (1964), 37–56].
3. M. A. Akcoglu, A pointwise ergodic theorem in L_p-spaces, *Canad. J. Math.* (to appear).
4. R. L. Adler, A. G. Konheim, and M. H. McAndrew, Topological entropy, *Trans. Amer. Math. Soc.* **114** (1965), 309–319.
5. W. Ambrose, Representation of ergodic flows, *Ann. of Math.* (2) **42** (1941), 723–739.
6. K. R. Berg, Convolution of invariant measures, maximal entropy, *Math. Systems Theory* **3** (1969), 146–150.
7. P. Billingsley, "Ergodic Theory and Information." Wiley, New York, 1965.
8. J. Blum and D. Hanson, On the isomorphism problem for Bernoulli schemes, *Bull. Amer. Math. Soc.* **69** (1963), 221–223.
9. L. Breiman, The individual ergodic theorem of information theory, *Ann. Math. Statist.* **28** (1957), 809–811; Correction in **31** (1960), 809–810.
10. L. Breiman, On achieving channel capacity in finite-memory channels, *Illinois J. Math.* **4** (1960), 246–252.
11. J. R. Brown, A universal model for dynamical systems with quasi-discrete spectrum, *Bull. Amer. Math. Soc.* **75** (1969), 1028–1030.
12. J. R. Brown, Inverse limits, entropy and weak isomorphism for discrete dynamical systems, *Trans. Amer. Math. Soc.* **164** (1972), 55–66.
13. J. R. Brown, A model for ergodic automorphisms on groups, *Math. Systems Theory* **6** (1972), 235–240.

14. J. R. Choksi, Inverse limits of measure spaces, *Proc. London Math. Soc.* (3) **8** (1958), 321–342.
15. E. I. Dinaburg, The relation between topological entropy and metric entropy, *Dokl. Akad. Nauk SSSR* **190** (1970), 19–22 [*Soviet Math. Dokl.* **11** (1970), 13–16].
16. N. Dunford and J. T. Schwartz, "Linear Operators," Part I. Wiley (Interscience), New York, 1958.
17. R. Ellis, Locally compact transformation groups, *Duke Math. J.* **24** (1957), 119–125.
18. R. Ellis, "Lectures on Topological Dynamics." Benjamin, New York, 1969.
19. A. Feinstein, "Foundations of Information Theory." McGraw-Hill, New York, 1958.
20. N. A. Friedman and D. S. Ornstein, Ergodic transformations induce mixing transformations, *Advances in Math.* **10** (1973), 147–163.
21. H. Furstenberg, Strict ergodicity and transformation of the torus, *Amer. J. Math.* **83** (1961), 573–601.
22. H. Furstenberg, The structure of distal flows, *Amer. J. Math.* **85** (1963), 477–515.
23. H. Furstenberg, Disjointness in ergodic theory, minimal sets, and a problem in Diophantine approximation, *Math. Systems Theory* **1** (1967), 1–49.
24. A. Garsia, "Topics in Almost Everywhere Convergence." Markham, Chicago, 1970.
25. T. N. T. Goodman, Relating topological entropy and measure entropy, *Bull. London Math. Soc.* **3** (1971), 176–180.
26. L. W. Goodwyn, Topological entropy bounds measure-theoretic entropy, *Proc. Amer. Math. Soc.* **23** (1969), 679–688.
27. L. W. Goodwyn, Some counter-examples in topological entropy, *Topology* **11** (1972), 377–385.
28. W. H. Gottschalk and G. A. Hedlund, Topological dynamics, *Amer. Math. Soc. Colloq. Publ.* **36**, Providence, 1955.
29. F. J. Hahn and W. Parry, Minimal dynamical systems with quasi-discrete spectrum, *J. London Math. Soc.* **40** (1965), 309–323.
30. M. Hall, "Combinatorial Theory." Blaisdell, Waltham, Massachusetts, 1967.
31. P. R. Halmos, "Measure Theory." Van Nostrand, Princeton, New Jersey, 1950.
32. P. R. Halmos, "Lectures on Ergodic Theory." Publ. Math. Soc. Japan, No. 3, Tokyo, 1956.
33. P. R. Halmos, Entropy in Ergodic Theory. Univ. of Chicago Lecture Notes, 1959.
34. P. R. Halmos and J. von Neumann, Operator methods in classical mechanics, II, *Ann. of Math.* Ser. II, **43** (1942), 332–350.
35. K. Jacobs, Ergodic decomposition of the Kolmogorov–Sinaĭ invariant, *in* "Ergodic Theory," F. B. Wright (ed.), pp. 173–190. Academic Press, New York, 1963.
36. S. Kakutani, Induced measure preserving transformations, *Proc. Imp. Acad. Tokyo* (*Japan Acad.*) **19** (1943), 635–641.
37. S. Kakutani, Examples of ergodic measure preserving transformations which are weakly mixing but not strongly mixing, *in* "Recent Advances in Topological Dynamics," A. Beck (ed.). Springer, New York, 1973.
38. I. Kaplansky, "Infinite Abelian Groups." Univ. of Michigan Press, Ann Arbor, 1954.
39. J. D. Kerrick, Group automorphisms of the N-torus: a representation theorem and some applications. Ph.D. dissertation, Oregon State Univ., Corvallis, 1972.
40. H. B. Keynes and J. B. Robertson, Generators for topological entropy and expansiveness, *Math. Systems Theory* **3** (1969), 51–59.
41. A. N. Kolmogorov, A new metric invariant of transient dynamical systems and automorphisms of Lebesgue spaces, *Dokl. Akad. Nauk. SSSR* **119** (1958), 861–864. (In Russian.)
42. W. Krieger, On unique ergodicity, *Proc. Sixth Berkeley Symp. Math. Statist. and Probability*, Vol. II, pp. 327–346. Univ. of California Press, Berkeley, 1972.
43. B. McMillan, The basic theorems of information theory, *Ann. Math. Statist.* **24** (1953), 196–219.

44. L. D. Meshalkin, A case of isomorphism of Bernoulli schemes, *Dokl. Akad. Nauk SSSR* **128** (1959), 41–44. (In Russian.)
45. D. S. Ornstein, Bernoulli shifts with the same entropy are isomorphic, *Advances in Math.* **4** (1970), 337–352.
46. D. S. Ornstein, Two Bernoulli shifts with infinite entropy are isomorphic, *Advances in Math.* **5** (1970), 339–348.
47. D. S. Ornstein, Imbedding Bernoulli shifts in flows, *Springer Lecture Notes* **160** (1970), 178–218.
48. D. S. Ornstein, An example of a Kolmogorov automorphism that is not a Bernoulli shift, *Advances in Math.* **10** (1973), 49–62.
49. J. C. Oxtoby, Ergodic sets, *Bull. Amer. Math. Soc.* **58** (1952), 116–136.
50. W. L. Reddy, Lifting expansive homeomorphisms to symbolic flows, *Math. Systems Theory* **2** (1968), 91–92.
51. V. A. Rohlin, Exact endomorphisms of a Lebesgue space, *Izv. Akad. Nauk SSSR Ser. Mat.* **25** (1961), 499–530 [*Amer. Math. Soc. Transl.* Ser. II, **39** (1963), 1–36.
52. V. A. Rohlin, On the entropy of automorphisms of a compact commutative group, *Theor. Probability Appl.* **6** (1961), 322–323.
53. W. Rudin, "Fourier Analysis on Groups." Wiley (Interscience), New York, 1962.
54. A. Saleski, On induced transformations of Bernoulli shifts, *Math. Systems Theory* **7** (1973), 83–96.
55. H. Schubert, "Topology." Allyn and Bacon, Boston, 1968.
56. T. L. Seethoff, Zero-entropy automorphisms of a compact abelian group, Tech. Report No. 40, Oregon State University Department of Mathematics, Corvallis, 1968.
57. P. Shields, "The Theory of Bernoulli Shifts." Univ. of Chicago Press, Chicago, 1973.
58. Ya. G. Sinaĭ, On the concept of entropy for dynamical systems, *Dokl. Akad. Nauk SSSR* **124** (1959), 768–771. (In Russian.)
59. M. Smorodinsky, Ergodic theory, entropy, *Springer Lecture Notes* **214** (1970).
60. M. D. Weiss, Algebraic and other entropies of group endomorphisms, *Math. Systems Theory* **8** (1975), 243–248.

ADDITIONAL REFERENCES

61. L. M. Abramov and V. A. Rohlin, Entropy of a skew product of transformation with invariant measure, *Vestnik Leningrad. Univ.* **7** (1962), 5–13. (In Russian.)
62. R. L. Adler and B. Weiss, Entropy, a complete metric invariant for automorphisms of the torus, *Proc. Nat. Acad. Sci. US* **57** (1967), 1573–1576.
63. H. Anzai and S. Kakutani, Bohr compactifications of a locally compact abelian group I & II, *Proc. Imp. Acad. Tokyo* (*Japan Acad.*) **19** (1943), 476–480, 533–539.
64. A. Beck and J. T. Schwartz, A vector-valued random ergodic theorem, *Proc. Amer. Math. Soc.* **8** (1957), 1049–1059.
65. A. Brunel and M. Keane, Ergodic theorems for operator sequences, *Z. Wahrschein. verw. Geb.* **12** (1969), 231–240.
66. R. V. Chacon, Identification of the limit of operator averages, *J. Math. Mech.* **11** (1962), 961–968.
67. R. Ellis, Distal transformation groups, *Pacific J. Math.* **8** (1958), 401–405.
68. R. Ellis, A semigroup associated with a transformation group, *Trans. Amer. Math. Soc.* **94** (1960), 272–281.
69. R. Ellis and W. H. Gottschalk, Homomorphisms of transformation groups, *Trans. Amer. Math. Soc.* **94** (1966), 258–271.

70. N. A. Friedman, "Introduction to Ergodic Theory." Van Nostrand-Reinhold, Princeton, New Jersey, 1970.
71. N. A. Friedman, Bernoulli shifts induce Bernoulli shifts, *Advances in Math.* **10** (1973), 39–48.
72. W. H. Gottschalk, Minimal sets: an introduction to topological dynamics, *Bull. Amer. Math. Soc.* **64** (1958), 336–351.
73. F. J. Hahn, On affine transformations of compact abelian groups, *Amer. J. Math.* (3) **85** (1963), 428–446.
74. F. Hahn and Y. Katznelson, On the entropy of uniquely ergodic transformations, *Trans. Amer. Math. Soc.* **126** (1967), 335–360.
75. F. Hahn and W. Parry, Some characteristic properties of dynamical systems with quasi-discrete spectra, *Math. Systems Theory* **2** (1968), 179–190.
76. P. R. Halmos and H. Samelson, On monothetic groups, *Proc. Nat. Acad. Sci. US* **28** (1942), 254–258.
77. D. L. Hanson and G. Pledger, On the mean ergodic theorem for weighted averages, *Z. Wahrschein. verw. Geb.* **13** (1969), 141–149.
78. G. A. Hedlund, Endomorphisms and automorphisms of the shift dynamical system, *Math. Systems Theory* **3** (1969), 320–375.
79. A. H. M. Hoare and W. Parry, Semi-groups of affine transformations, *Quart. J. Math. Oxford* (2) **17** (1966), 106–111.
80. A. H. M. Hoare and W. Parry, Affine transformations with quasi-discrete spectrum (I), *J. London Math. Soc.* **41** (1966), 88–96.
81. E. Hopf, "Ergodentheorie." Springer, Berlin, 1937.
82. S. A. Juzvinskiĭ, Metric properties of endomorphisms of compact groups, *Izv. Akad. Nauk. SSSR Ser. Mat.* **29** (1965), 1295–1328 [*Amer. Math. Soc. Transl.* Ser. 2, **66** (1966), 63–98].
83. S. Kakutani, Random ergodic theorems and Markoff processes with a stable distribution, *Proc. Second Berkeley Symp. Probability and Statist.*, pp. 247–261. Univ. of California Press, Berkeley, 1951.
84. S. Kakutani, Determination of the spectrum of the flow of Brownian motion, *Proc. Nat. Acad. Sci. US* **36** (1950), 319–323.
85. S. Kakutani, Ergodic theory, *Proc. Intern. Congress of Mathematicians*, pp. 319–323. Cambridge, 1952.
86. S. Kakutani and W. Parry, Infinite measure preserving transformation with "mixing," *Bull. Amer. Math. Soc.* **69** (1963), 752–756.
87. Y. Katznelson, Ergodic automorphisms of T^n are Bernoulli shifts, *Israel J. Math.* **10** (1971), 186–195.
88. U. Krengel, Entropy of conservative transformations, *Z. Wahrschein. verw. Geb.* **7** (1967), 161–181.
89. P. -F. Lam, On expansive transformation groups, *Trans. Amer. Math. Soc.* **150** (1970), 131–138.
90. D. Maharam, On orbits under ergodic measure-preserving transformations, *Trans. Amer. Math. Soc.* **119** (1965), 51–66.
91. J. Neveu, Une demonstration simplifiee et une extension de la formule d'Abramov sur l'entropie des transformations induites, *Z. Wahrschein. verw. Geb.* **13** (1969), 135–140.
92. D. S. Ornstein, On invariant measures, *Bull. Amer. Math. Soc.* **66** (1960), 297–300.
93. D. S. Ornstein, A K-automorphism with no square root and Pinsker's conjecture, *Advances in Math.* **10** (1973), 89–102.
94. D. S. Ornstein, A mixing transformation for which Pinsker's conjecture fails, *Advances in Math.* **10** (1973), 103–123.

95. D. S. Ornstein, The isomorphism theorem for Bernoulli flows, *Advances in Math.* **10** (1973), 124–142.
96. D. S. Ornstein and P. C. Shields, An uncountable family of K-automorphisms, *Advances in Math.* **10** (1973), 63–88.
97. D. S. Ornstein and P. C. Shields, Mixing Markov shifts of kernel type are Bernoulli, *Advances in Math.* **10** (1973), 143–146.
98. W. Parry, Intrinsic Markov chain, *Trans. Amer. Math. Soc.* **112** (1964), 55–56.
99. W. Parry, On the coincidence of three invariant σ-algebras associated with an affine transformation, *Proc. Amer. Math. Soc.* **17** (1966), 1297–1302.
100. W. Parry, Entropy and Generators in Ergodic Theory. Lecture Notes, Yale University Department of Mathematics, New Haven, 1966.
101. W. Parry and P. Walters, Minimal skew product homeomorphisms and coalescence, *Compositio Math.* **22** (1970), 283–288.
102. V. A. Rohlin, Selected topics from the metric theory of dynamical systems, *Uspehi Mat. Nauk* **4** (1949), 57–128 [*Amer. Math. Soc. Transl.* Ser. I, **49** (1960), 171–240].
103. V. A. Rohlin, Metric properties of endomorphisms of compact commutative groups, *Izv. Akad. Nauk SSSR Ser. Mat.* **28** (1964), 867–874. (In Russian.)
104. G.-C. Rota, On the maximal ergodic theorem for Abel limits, *Proc. Amer. Math. Soc.* **14** (1963), 722–723.
105. M. Sears, The automorphisms of the shift dynamical system are relatively sparse, *Math. Systems Theory* **5** (1971), 228–231.
106. P. C. Shields, Cutting and independent stacking of intervals, *Math. Systems Theory* **10** (1973), 1–4.
107. Ya. G. Sinaĭ, Probabilistic ideas in ergodic theory, *Amer. Math. Soc. Transl.* (2) **31** (1963), 62–84.
108. Ya. G. Sinaĭ, Weak isomorphism of transformations with invariant measure, *Mat. Sb.* (N.S.) **63** (105) (1964), 23–42 [*Amer. Math. Soc. Transl.* Ser. 2, **57** (1966), 123–143].
109. M. Smorodinsky, A partition on a Bernoulli shift which is not weakly Bernoulli, *Math. Systems Theory* **5** (1971), 201–203.
110. M. Smorodinsky, On Ornstein's isomorphism theorem for Bernoulli shifts, *Advances in Math.* **9** (1972), 1–9.
111. P. Walters, On the relationship between zero entropy and quasi-discrete spectrum for affine transformations, *Proc. Amer. Math. Soc.* **18** (1967), 661–667.
112. P. Walters, Topological conjugacy of affine transformations of compact abelian groups, *Trans. Amer. Math. Soc.* **140** (1969), 95–107.
113. P. Walters, Conjugacy properties of affine transformations of nilmanifolds, *Math. Systems Theory* **4** (1970), 327–333.
114. P. Walters, Some invariant σ-algebras for measure-preserving transformations, *Trans. Amer. Math. Soc.* **163** (1972), 357–368.
115. B. Weiss, The isomorphism problem in ergodic theory, *Bull. Amer. Math. Soc.* **78** (1972), 668–684.

Index

K

L

M

N

O

P

Q

R

S

T

U

V

W

Y

A 6
B 7
C 8
D 9
E 0
F 1
G 2
H 3
I 4
J 5